I0010762

Google Cloud for DevOps Engineers

A practical guide to SRE and achieving Google's
Professional Cloud DevOps Engineer certification

Sandeep Madamanchi

BIRMINGHAM—MUMBAI

Google Cloud for DevOps Engineers

Copyright © 2021 Packt Publishing

All rights reserved. No part of this book may be reproduced, stored in a retrieval system, or transmitted in any form or by any means, without the prior written permission of the publisher, except in the case of brief quotations embedded in critical articles or reviews.

Every effort has been made in the preparation of this book to ensure the accuracy of the information presented. However, the information contained in this book is sold without warranty, either express or implied. Neither the author, nor Packt Publishing or its dealers and distributors, will be held liable for any damages caused or alleged to have been caused directly or indirectly by this book.

Packt Publishing has endeavored to provide trademark information about all of the companies and products mentioned in this book by the appropriate use of capitals. However, Packt Publishing cannot guarantee the accuracy of this information.

Group Product Manager: Wilson D'souza

Publishing Product Manager: Vijin Boricha

Senior Editor: Rahul D'souza

Content Development Editor: Nihar Kapadia

Technical Editor: Sarvesh Jaywant

Copy Editor: Safis Editing

Project Coordinator: Neil D'mello

Proofreader: Safis Editing

Indexer: Vinayak Purushotham

Production Designer: Nilesh Mohite

First published: July 2021

Production reference: 3280621

Published by Packt Publishing Ltd.

Livery Place

35 Livery Street

Birmingham

B3 2PB, UK.

ISBN 978-1-83921-801-9

www.packt.com

To my wonderful daughter – Adwita. As I see you grow every day,
you motivate me to learn and inspire me to be a better person.
Thank You!!

Contributors

About the author

Sandeep Madamanchi works at a company called Variant that re-engineers trucking through technology, acting as the Head of *Cloud Infrastructure and Data Engineering*. He is a continuous learner, focused on building highly resilient, secure, and self-healing cloud infrastructure. He advocates SRE and its practices to achieve reliability and operational stability. His vision is to provide infrastructure as a service to core engineering, analytics, and ML specialized teams through DevOps, DataOps, and MLOps practices.

Prior to Variant, he was Director of R&D at Manhattan Associates; a leader in supply chain software and worked there for over 14 years. He holds certifications across *AWS* and *GCP*; which includes the Professional *Cloud DevOps Engineer Certification*.

I want to thank my family, friends, and mentors. A big shout out to the Packt team (Neil, Nihar, Rahul, Riyan) led by Vijin Boricha for their professionalism, and to the technical reviewers for their constructive feedback. A big "thank you" to my best friend – Mridula, who kickstarted my cloud journey. Finally, I would like to express my gratitude to my parents (both of whom are educators) and to my grandparents for their support during my learning phase.

About the reviewers

Richard Rose has worked in IT for over 20 years and holds many industry certifications. He is the author of *Hands-On Serverless Computing with Google Cloud*. As someone who works with the cloud every day, Rich is a strong proponent of cloud certifications. For him, gaining expertise in an area is not the end of the journey. It is the start of realizing your potential.

In his spare time, he enjoys spending time with his family and playing the guitar. Recently, he has updated his blog to include his development-side projects and general computing tips.

I would like to acknowledge the time, support, and energy of Dawn, Bailey, Elliot, Noah, and Amelia. Without your assistance, it would be impossible to achieve the possible.

Vaibhav Chopra has extensive experience in the fields of DevOps, SRE, Cloud Infra, and NFV. He currently works as a senior engineering manager for Orange, and prior to that, he worked with Ericsson, Amdocs, and Netcracker over a period of 11 years.

He is an open source enthusiast and is passionate about the latest technologies, as well as being a multi-cloud expert and a seasoned blogger.

LFN has awarded him for his contribution in open source projects and he has participated as a trainer and speaker at multiple conferences. In his current role, he drives the team in relation to various CNCF projects, largely revolving around container security, resiliency, and benchmarking, to build a production-grade hybrid cloud (GCP Anthos and Openstack) solution.

I would like to thank 3 ladies, my mother Anu, my wife, Swati, my daughter, Saanvika, who have supported me to contribute extra time towards this and to my friends, who continuously motivate and inspire me to learn new things.

Bruno S. Brasil is a cloud engineer who has always used Linux and then started working in on-premises environments, participating in the modernization and migration to cloud solutions. After working with several cloud providers, he chose Google Cloud as his focus of expertise. Since then, he has worked on projects of this type as a consultant and engineer, for several types of businesses, ranging from digital banks and marketplaces to start-ups. He has always focused on implementing best practices in the development of infrastructure as code, disseminating the DevOps culture and implementing SRE strategies. He is enthusiastic about the open source community and believes that this is the most important path in terms of the growth of new professionals.

I would like to thank my family and friends who have always supported me since I chose to follow the path of technology. It was difficult to begin with, but the fact that I am here today is because of the support I have received from them. I would also like to thank the open source community, everyone who shares their experiences, solutions, and time, for organizing events of all kinds. This is essential to making knowledge accessible and democratic.

Table of Contents

2

SRE Technical Practices – Deep Dive

3

Understanding Monitoring and Alerting to Target Reliability

4
Building SRE Teams and Applying Cultural Practices

Section 2: Google Cloud Services to Implement DevOps via CI/CD

5
Managing Source Code Using Cloud Source Repositories

6

Building Code Using Cloud Build, and Pushing to Container Registry

7

Understanding Kubernetes Essentials to Deploy Containerized Applications

8

Understanding GKE Essentials to Deploy Containerized Applications

9

Securing the Cluster Using GKE Security Constructs

Appendix

Getting Ready for Professional Cloud DevOps Engineer Certification

Mock Exam 1

Mock Exam 2

Other Books You May Enjoy

Index

Preface

This book is a comprehensive guide to **Site Reliability Engineering** (**SRE**) fundamentals, Google's approach to DevOps. In addition, the book dives into critical services from Google Cloud to implement **Contiuous Integration/Continous Deployment** (**CI/CD**) with a focus on containerized deployments via Kubernetes. The book also serves as preparation material for the Professional Cloud DevOps Engineer certification from Google Cloud with chapter-based tests and mock tests included.

Who this book is for

This book is ideal for cloud system administrators and network engineers interested in resolving cloud-based operational issues. IT professionals looking to enhance their careers in administering Google Cloud services will benefit. Users who want to learn about applying SRE principles and are focused on implementing DevOps in **Google Cloud Platform** (**GCP**) will also benefit. Basic knowledge of cloud computing and GCP services, an understanding of CI/CD, and hands-on experience with Unix/Linux infrastructure are recommended. Those interested in passing the Professional Cloud DevOps Engineer certification will find this book useful.

The Professional Cloud DevOps Engineer certification is advanced in nature. To better prepare for this certification, it is recommended to either prepare for or be certified in the Associate Cloud Engineer certification or Professional Cloud Architect certification. Though these certifications are not prerequisites for the Professional Cloud DevOps Engineer certification, they help to better prepare and be acquainted with services on GCP.

What this book covers

Chapter 1, DevOps, SRE, and Google Cloud Services for CI/CD, covers DevOps, which is a set of practices that builds, tests, and releases code in a repeatable and iterative manner. These practices are aimed to break down the metaphoric wall between development and operation teams. SRE is a prescriptive way from Google to implement DevOps that aligns incentives between the development and operations teams to build and maintain reliable systems. In addition, Google recommends a cloud-native development paradigm where complex systems are decomposed into multiple services using microservices architecture.

This chapter will cover topics that include the DevOps life cycle, the evolution of SRE, an introduction to key technical and cultural SRE practices, and the benefits of using a cloud-native development paradigm. The chapter will also introduce services on GCP to implement cloud-native development and apply SRE concepts.

Chapter 2, SRE Technical Practices – Deep Dive, covers reliability, which is the most critical feature of a service and should be aligned with business objectives. SRE prescribes specific technical practices to measure characteristics that define and track reliability. These technical practices include the **Service-Level Agreement (SLA)**, **Service-Level Objective (SLO)**, **Service-Level Indicator (SLI)**, error budgets, and eliminating toil through automation.

This chapter goes in depth into SRE technical practices. This chapter will cover topics that include the blueprint for a well-defined SLA, defining reliability expectations via SLOs, understanding reliability targets and their implications, categorizing user journeys, sources to measure SLIs, exploring ways to make a service reliable by tracking error budgets, and eliminating toil through automation. The chapter concludes by walking through two scenarios where the impact of SLAs, SLOs, and error budgets are illustrated relative to the SLI being measured.

Chapter 3, Understanding Monitoring and Alerting to Target Reliability, discusses how the key to implementing SRE technical practices is to ensure that SLAs, SLOs, and SLIs are never violated. This makes it feasible to balance new feature releases and still maintain system reliability since the error budget is not exhausted. Monitoring, alerting, and time series are fundamental concepts to track SRE technical practices.

This chapter goes in depth into monitoring, alerting, and time series. This chapter will cover topics that include monitoring sources, monitoring types, monitoring strategies, the golden signals that are recommended to be measured and monitored, potential approaches and key attributes to define an alerting strategy, and the structure and cardinality of time series.

Chapter 4, Building SRE Teams and Applying Cultural Practices, covers how for a long time, Google considered SRE as their secret sauce to achieving system reliability while maintaining a balance with new feature release velocity. Google achieved this by applying a prescribed set of cultural practices such as incident management, being on call, and psychological safety. SRE cultural practices are required to implement SRE technical practices and are strongly recommended by Google for organizations that would like to start their SRE journey. In addition, Google has put forward aspects that are critical to building SRE teams along with an engagement model.

This chapter goes in depth into building SRE teams, which includes topics around different implementations of SRE teams, details around staffing SRE engineers, and insights into the SRE engagement model. This chapter will also cover topics on SRE cultural practices that include facets of effective incident management, factors to consider while being on call, and factors to overcome to foster psychological safety. The chapter concludes with a cultural practice that is aimed to reduce organizational silos by sharing vision and knowledge and by fostering collaboration.

Chapter 5, Managing Source Code Using Cloud Source Repositories, looks at how source code management is the first step in a CI flow. Code is stored in a source code repository such as GitHub or Bitbucket so that developers can continuously make code changes and the modified code is integrated into the repository. **Cloud Source Repositories** (CSR) is a service from Google Cloud that provides source code management through private Git repositories.

This chapter goes in depth into CSR. This chapter will cover topics that include key features of CSR, steps to create and access the repository, how to perform one-way sync from GitHub/Bitbucket to CSR, and common operations in CSR such as browsing repositories, performing universal code search, and detecting security keys. The chapter concludes with a hands-on lab that illustrates how code can be deployed in Cloud Functions by pulling code hosted from CSR.

Chapter 6, Building Code Using Cloud Build, and Pushing to Container Registry, looks at how once code is checked into a source code management system such as CSR, the next logical step in a CI flow is to build code, create artifacts, and push to a registry that can store the generated artifacts. Cloud Build is a service from Google Cloud that can build the source code whereas Container Registry is the destination where the created build artifacts are stored.

This chapter goes in depth into Cloud Build and Container Registry. This chapter will cover topics that include understanding the need for automation, processes to build and create container images, key essentials of Cloud Build, strategies to optimize the build speed, key essentials of Container Registry, the structure of Container Registry, and Container Analysis. The chapter concludes with a hands-on lab to build, create, push, and deploy a container to Cloud Run using Cloud Build triggers. This hands-on lab also illustrates a way to build a CI/CD pipeline as it includes both CI and automated deployment of containers to Cloud Run, a GCP compute option that runs containers.

Chapter 7, Understanding Kubernetes Essentials to Deploy Containerized Applications, covers **Kubernetes**, or **K8s**, which is an open source container orchestration system that can run containerized applications but requires significant effort in terms of setting up and ongoing maintenance. Kubernetes originated as an internal cluster management tool from Google; Google donated it to the **Cloud Native Computing Foundation** (**CNCF**) as an open source project in 2014.

This chapter goes in depth into K8s. This chapter will cover topics that include key features of K8s, elaboration of cluster anatomy, which includes components of the master control plane, node components, key Kubernetes objects such as Pods, Deployments, StatefulSets, DaemonSet, Job, CronJob, and Services, and critical factors that need to be considered while scheduling Pods. The chapter concludes with a deep dive into possible deployment strategies in Kubernetes, which includes Recreate, Rolling Update, Blue/Green, and Canary.

Chapter 8, Understanding GKE Essentials to Deploy Containerized Applications, covers **Google Kubernetes Engine** (**GKE**), which is a managed version of K8s; that is, an open source container orchestration system to automate application deployment, scaling, and cluster management. GKE requies less effort in terms of cluster creation and ongoing maintenance.

This chapter goes in depth into GKE. This chapter will cover topics that include GKE core features such as GKE node pools, GKE cluster configurations, GKE autoscaling, networking in GKE, which includes Pod and service networking, GKE storage options, and Cloud Operations for GKE. There are two hands-on labs in this chapter. The first hands-on lab is placed at the start of the chapter and illustrates cluster creation using Standard mode, deploying workloads, and exposing the Pod as a service. The chapter concludes with another hands-on lab like the first one but with the cluster creation mode as Autopilot.

Chapter 9, Securing the Cluster Using GKE Security Constructs, covers how securing a Kubernetes cluster is a critical part of deployment. Native Kubernetes provide some essential security features that focus on how a request being sent to the cluster is authenticated and authorized. It is also important to understand how the master plane components are secured along with the Pods running the applications. Additionally, GKE provides security features that are fundamental to harden a cluster's security.

This chapter goes in depth into GKE security features. This chapter will cover topics that include essential security patterns in Kubernetes, control plane security, and Pod security. This chapter concludes by discussing various GKE-specific security features such as GKE private clusters, container-optimized OS, shielded GKE nodes, restricting traffic among Pods using a network policy, deploying time security services via binary authorizations, and using a workload identity to access GCP services from applications running inside the GKE cluster.

Chapter 10, *Exploring GCP Cloud Operations*, looks at how once an application is deployed, the next critical phase of the DevOps life cycle is continuous monitoring as it provides a feedback loop to ensure the reliability of a service or a system. As previously discussed in *Chapter 2*, *SRE Technical Practices – Deep Dive*, SRE prescribes specific technical tools or practices that help in measuring characteristics that define and track reliability, such as SLAs, SLOs, SLIs, and error budgets. SRE prescribes the use of observability to track technical practices. Observability on GCP is established through Cloud Operations.

This chapter goes in depth into Cloud Operations. This chapter will cover topics that include Cloud Monitoring essentials such as workspaces, dashboards, Metrics Explorer, uptime checks, configuring alerting, and the need for a monitoring agent and access controls specific to Cloud Monitoring. The chapter will also cover topics that include Cloud Logging essentials such as audit logs with their classification, summarizing log characteristics across log buckets, logs-based metrics, access controls specific to Cloud Logging, network-based log types, and the use of logging agents. This chapter concludes by discussing various essentials tied to Cloud Debugger, Cloud Trace, and Cloud Profiler.

To get the most out of this book

It is recommended that you have prior knowledge on topics that include Docker, an introduction to native Kubernetes, a working knowledge of Git, getting hands on with key services on Google Cloud such as Cloud Operations, compute services, and hands-on usage of Google Cloud SDK or Cloud Shell. Additionally, hands-on knowledge of a programming language of choice such as Python, Java, or Node.js will be very useful. The code samples in this book are, however, written in Python.

Software/hardware covered in the book	OS requirements
Python 3.7 or above	Windows, macOS, or Linux (any)
Google Cloud SDK	`https://cloud.google.com/sdk/docs/quickstart`

If you are using the digital version of this book, we advise you to type the code yourself or access the code via the GitHub repository (link available in the next section). Doing so will help you avoid any potential errors related to the copying and pasting of code.

Irrespective of whether you are working toward the Professional Cloud DevOps Engineer certification, it is recommended to attempt the mock exam after completing all the chapters. This will be a good way to assess the learnings absorbed from the content of the book.

Download the example code files

The code bundle for the book is also hosted on GitHub at `https://github.com/PacktPublishing/Google-Cloud-Platform-for-DevOps-Engineers`. In case there's an update to the code, it will be updated on the existing GitHub repository.

We also have other code bundles from our rich catalog of books and videos available at `https://github.com/PacktPublishing/`. Check them out!

Download the color images

We also provide a PDF file that has color images of the screenshots/diagrams used in this book. You can download it here: `http://www.packtpub.com/sites/default/files/downloads/9781839218019_ColorImages.pdf`.

Conventions used

There are a number of text conventions used throughout this book.

`Code in text`: Indicates code words in text, database table names, folder names, filenames, file extensions, pathnames, dummy URLs, user input, and Twitter handles. Here is an example: "You can use the `my-first-csr` repository."

A block of code is set as follows:

```
steps:
- name: 'gcr.io/cloud-builders/docker'
  args: ['build', '-t', 'gcr.io/$PROJECT_ID/builder-myimage',
'.']
- name: 'gcr.io/cloud-builders/docker'
  args: ['push', 'gcr.io/$PROJECT_ID/builder-myimage']
- name: 'gcr.io/cloud-builders/gcloud'
```

When we wish to draw your attention to a particular part of a code block, the relevant lines or items are set in bold:

```
apiVersion: autoscaling.k8s.io/v1
kind: VerticalPodAutoscaler
metadata:
  name: my-vpa
```

Any command-line input or output is written as follows:

```
gcloud builds submit --config <build-config-file> <source-code-
path>
```

Bold: Indicates a new term, an important word, or words that you see onscreen. For example, words in menus or dialog boxes appear in the text like this. Here is an example: "Navigate to **Source Repositories** within GCP and select the **Get Started** option."

> **Tips or important notes**
> Appear like this.

Get in touch

Feedback from our readers is always welcome.

General feedback: If you have questions about any aspect of this book, mention the book title in the subject of your message and email us at customercare@packtpub.com.

Errata: Although we have taken every care to ensure the accuracy of our content, mistakes do happen. If you have found a mistake in this book, we would be grateful if you would report this to us. Please visit www.packtpub.com/support/errata, selecting your book, clicking on the Errata Submission Form link, and entering the details.

Piracy: If you come across any illegal copies of our works in any form on the Internet, we would be grateful if you would provide us with the location address or website name. Please contact us at copyright@packt.com with a link to the material.

If you are interested in becoming an author: If there is a topic that you have expertise in and you are interested in either writing or contributing to a book, please visit authors.packtpub.com.

Reviews

Please leave a review. Once you have read and used this book, why not leave a review on the site that you purchased it from? Potential readers can then see and use your unbiased opinion to make purchase decisions, we at Packt can understand what you think about our products, and our authors can see your feedback on their book. Thank you!

For more information about Packt, please visit `packt.com`.

Section 1: Site Reliability Engineering – A Prescriptive Way to Implement DevOps

The core focus of this section is **Site Reliability Engineering (SRE)**. The section starts with the evolution of DevOps and its life cycle and introduces SRE as a prescriptive way to implement DevOps. The emphasis is on defining the reliability of a service and ways to measure it. In this section, SRE technical practices such as **Service-Level Agreement (SLA)**, **Service-Level Objective (SLO)**, **Service-Level Indicator (SLI)**, error budget, and eliminating toil are introduced and are further elaborated. Fundamentals and critical concepts around monitoring, alerting, and time series are also explored to track SRE technical practices. However, SRE technical practices cannot be implemented within an organization without bringing cultural change. Google prescribes a set of SRE cultural practices such as incident management, being on call, psychological safety, and the need to foster collaboration. This section also explores the aspects critical to building SRE teams and provides insights into the SRE engagement model. It's important to note that the start of the section introduces services in Google Cloud to implement DevOps through the principles of SRE; however, the details of those services are explored in the next section.

This part of the book comprises the following chapters:

- *Chapter 1, DevOps, SRE, and Google Cloud Services for CI/CD*
- *Chapter 2, SRE Technical Practices – Deep Dive*
- *Chapter 3, Understanding Monitoring and Alerting to Target Reliability*
- *Chapter 4, Building SRE Teams and Applying Cultural Practices*

1
DevOps, SRE, and Google Cloud Services for CI/CD

DevOps is a mindset change that tries to balance release velocity with system reliability. It aims to increase an organization's ability to continuously deliver reliable applications and services at a high velocity when compared to traditional software development processes.

A common misconception about DevOps is that it is a technology. Instead, DevOps is a set of supporting practices (such as, build, test, and deployment) that combines software development and IT operations. These practices establish a culture that breaks down the metaphorical wall between developers (who aim to push new features to production) and system administrators or operators (who aim to keep the code running in production).

Site Reliability Engineering (SRE) is Google's approach to align incentives between development and operations that are key to building and maintaining reliable engineering systems. SRE is a prescriptive way to implement DevOps practices and principles. Through these practices, the aim is to increase overall observability and reduce the level of incidents. The introduction of a **Continuous Integration/Continuous Delivery (CI/CD)** pipeline enables a robust feedback loop in support of key SRE definitions such as toil, observability, and incident management.

CI/CD is a key DevOps practice that helps to achieve this mindset change. CI/CD requires a strong emphasis on automation to *build reliable software faster* (in terms of delivering/deploying to production). Software delivery of this type requires agility, which is often achieved by breaking down existing components.

A **cloud-native development** paradigm is one where complex systems are decomposed into multiple services (such as microservices architecture). Each service can be independently tested and deployed into an isolated runtime. **Google Cloud Platform (GCP)** has well-defined services to implement cloud-native development and apply SRE concepts to achieve the goal of *building reliable software faster*.

In this chapter, we're going to cover the following main topics:

- DevOps 101 – evolution and life cycle
- SRE 101 – evolution; technical and cultural practices
- GCP's cloud-native approach to implementing DevOps

Understanding DevOps, its evolution, and life cycle

This section focuses on the evolution of DevOps and lists phases or critical practices that form the DevOps life cycle.

Revisiting DevOps evolution

Let's take a step back and think about how DevOps has evolved. **Agile software development methodology** refers to a set of practices based on iterative development where requirements and solutions are built through collaboration between cross-functional teams and end users. DevOps can be perceived as a logical extension of Agile. Some might even consider DevOps as an offspring of Agile. This is because DevOps starts where Agile logically stops. Let's explore what this means in detail.

Agile was introduced as a holistic approach for end-to-end software delivery. Its core principles are defined in the Agile Manifesto (`https://agilemanifesto.org/`), with specific emphasis on interaction with processes and tools, improving collaboration, incremental and iterative development, and flexibility in response to changes to a fixed plan. The initial Agile teams primarily had developers, but it quickly extended to product management, customers, and quality assurance. If we factor in the impact of the increased focus on iterative testing and user acceptance testing, the result is a new capacity to deliver software faster to production.

However, Agile methodology creates a new problem that has resulted in a need for a new evolution. Once software is delivered to production, the operations team are primarily focused on system stability and upkeep. At the same time, development teams continue to add new features to a delivered software to meet customers' dynamic needs and to keep up with the competition.

Operators were always cautious for the fear of introducing issues. Developers always insist on pushing changes since these were tested in their local setup, and developers always thought that it is the responsibility of the operators to ensure that the changes work in production. But from an operator's standpoint, they have little or no understanding of the code base. Similarly, developers have little or no understanding of the operational practices. So essentially, developers were focused on shipping new features faster and operators were focused on stability. This forced developers to move slower in pushing the new features out to production. This misalignment often caused tensions within an organization.

Patrick Debois, an IT consultant who was working on a large data center migration project in 2007, experienced similar challenges when trying to collaborate with developers and operators. He coined the term DevOps and later continued this movement with Andrew Shafer. They considered DevOps as an extension of Agile. In fact, when it came to naming their first Google group for DevOps, they called it **Agile System Administration**.

The DevOps movement enabled better communication between software development and IT operations and effectively led to improved software with continuity being the core theme across operating a stable environment, consistent delivery, improved collaboration, and enhanced operational practices with a focus on innovation. This led to the evolution of the DevOps life cycle, which is detailed in the upcoming sub-section.

DevOps life cycle

DevOps constitutes phases or practices that in their entirety form the DevOps life cycle. In this section, we'll look at these phases in detail, as shown in the following diagram:

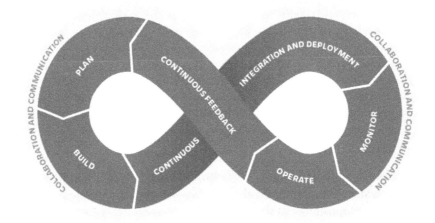

Figure 1.1 – Phases of the DevOps life cycle

There are six primary phases in a DevOps life cycle. They are as follows:

- Plan and build
- Continuous integration
- Continuous delivery
- Continuous deployment
- Continuous monitoring and operations
- Continuous feedback

The keyword here is **continuous**. If code is developed continuously, it will be followed with a need to continuously test, provide feedback, deploy, monitor, and operate. These phases will be introduced in the following sections.

Phase 1 – plan and build

In the **planning** phase, the core focus is to understand the vision and convert it into a detailed plan. The plan can be split into phases, otherwise known as **epics** (in Agile terminology). Each phase or epic can be scoped to achieve a specific set of functionalities, which could be further groomed as one or multiple user stories. This requires a lot of communication and collaboration between various stakeholders.

In the **build** phase, code is written in the language of choice and appropriate build artifacts are created. Code is maintained in a source code repository such as GitHub, Bitbucket, and others.

Phase 2 – continuous integration

CI is a software development practice where developers frequently integrate their code changes to the main branch of a shared repository. This is done, preferably, several times in a day, leading to several integrations.

> **Important note**
> **Code change** is considered the fundamental unit of software development. Since development is incremental in nature, developers keep changing their code.

Ideally, each integration is triggered by an automated build that also initiates automated unit tests, to detect any issues as quickly as possible. This avoids *integration hell*, or in other words, ensures that the application is not broken by introducing a code change or delta into the main branch.

Phase 3 – continuous delivery

Continuous delivery is a software development practice to build software such that a set of code changes can be delivered or released to production at any time. It can be considered an extension of CI and its core focus is on automating the release process to enable hands-free or single-click deployments.

The core purpose is to ensure that the code base is releasable and there is no regression break. It's possible that the newly added code might not necessarily work. The frequency to deliver code to production is very specific to the organization and could be daily, weekly, bi-weekly, and so on.

Phase 4 – continuous deployment

Continuous deployment is a software development practice where the core focus is to release automated deployments to production without the user's intervention. It aims to minimize the time elapsed between developers writing new line(s) of code and this new code being used by live users in production.

At its core, continuous deployment incorporates robust testing frameworks and encourages code deployment in a testing/staging environment post the continuous delivery phase. Automated tests can be run as part of the pipeline in the test/stage environment. In the event of no issues, the code can be deployed to production in an automated fashion. This removes the need for a formal release day and establishes a feedback loop to ensure that added features are useful to the end users.

Phase 5 – continuous monitoring and operation

Continuous monitoring is a practice that uses analytical information to identify issues with the application or its underlying infrastructure. Monitoring can be classified into two types: **server monitoring** and **application monitoring**.

Continuous operations is a practice where the core focus is to mitigate, reduce, or eliminate the impact of planned downtime, such as scheduled maintenance, or in the case of unplanned downtime, such as an incident.

Phase 6 – continuous feedback

Continuous feedback is a practice where the core focus is to collect feedback that improves the applica/service. A common misconception is that continuous feedback happens only as the last phase of the DevOps cycle.

Feedback loops are present at every phase of the DevOps pipeline such that feedback is conveyed if a build fails due to a specific code check-in, a unit/integration test or functional test fails in a testing deployment, or an issue is found by the customer in production.

GitOps is one of the approaches to implement continuous feedback where a version control system has the capabilities to manage operational workflows, such as Kubernetes deployment. A failure at any point in the workflow can be tracked directly in the source control and that creates a direct feedback loop.

Key pillars of DevOps

DevOps can be categorized into five key pillars or areas:

- **Reduce organizational silos**: Bridge the gap between teams by encouraging them to work together toward a shared company vision. This reduces friction between teams and increases communication and collaboration.

- **Accept failure as normal**: In the *continuous* aspect of DevOps, failure is considered an opportunity to continuously improve. Systems/services are bound to fail, especially when more features are added to improve the service. Learning from failures mitigates reoccurrence. Fostering failure as the normal culture will make team members more forthcoming.

- **Implement gradual change**: Implementing gradual change falls in line with the continuous aspect of DevOps. Small, gradual changes are not only easier to review but in the event of an incident in production, it is easier to roll back and reduce the impact of the incident by going back to a last known working state.

- **Leverage tooling and automation**: Automation is key to implement the continuous aspect of CI/CD pipelines, which are critical to DevOps. It is important to identify manual work and automate it in a way that eventually increases speed and adds consistency to everyday processes.

- **Measure everything**: Measuring is a critical gauge for success. Monitoring is one way to measure and observe that helps to get important feedback to continuously improve the system.

This completes our introduction to DevOps where we discussed its evolution, life cycle phases, and key pillars. At the end of the day, DevOps is a set of practices. The next section introduces site reliability engineering, or SRE, which is essentially Google's practical approach to implementing DevOps key pillars.

SRE's evolution; technical and cultural practices

This section tracks back the evolution of SRE, defines SRE, discusses how SRE relates to DevOps by elaborating DevOps key pillars, details critical jargon, and introduces SRE's cultural practices.

The evolution of SRE

In the early 2000s, Google was building massive, complex systems to run their search and other critical services. Their main challenge was to *reliably* run their services. At the time, many companies historically had system administrators deploying software components as a service. The use of system administrators, otherwise known as the *sysadmin* approach, essentially focused on running the service by responding to events or updates as they occur. This means that if the service grew in traffic or complexity, there would be a corresponding increase in events and updates.

The sysadmin approach has its pitfalls, and these are represented by two categories of cost:

- **Direct costs**: Running a service with a team of system administrators included manual intervention. Manual intervention at scale is a major downside to change management and event handling. However, this manual approach was adopted by multiple organizations because there wasn't a recognized alternative

- **Indirect costs**: System administrators and developers widely differed in terms of their skills, the vocabulary used to describe situations, and incentives. Development teams always want to launch new features and their incentive is to drive adoption. System administrators or ops teams want to ensure that the service is running reliably and often with a thought process of *don't change something that is working*.

Google did not want to pursue a manual approach because at their scale and traffic, any increase in demand would make it impractical to scale. The desire to regularly push more features to their users would ultimately cause conflict between developers and operators. Google wanted to reduce this conflict and remove the confusion with respect to desired outcomes. With this knowledge, Google considered an alternative approach. This new approach is what became known as SRE.

Understanding SRE

> *SRE is what happens when you ask a software engineer to design an operations team.*

(*Betsy Beyer, Chris Jones, Jennifer Petoff, & Niall Murphy, Site Reliability Engineering, O'REILLY*)

The preceding is a quote from Ben Treynor Sloss, who in 2003 started the first SRE team at Google with seven software engineers. Ben himself was a software engineer up until that point, and joined Google as the site reliability Tsar in 2003, led the development and operations of Google's production software infrastructure, network, and user-facing services, and is currently the VP of engineering at Google. At that point in 2003, neither Ben nor Google had any formal definition for SRE.

SRE is a software engineering approach to IT operations. SRE is an intrinsic part of Google's culture. It's the key to running their massively complex systems and services at scale. At its core, the goal of SRE is to end the age-old battle between development and operations. This section introduces SRE's thought process and the upcoming chapters on SRE give deeper insights into how SRE achieves its goal.

A primary difference in Google's approach to building the SRE practice or team is the composition of the SRE team. A typical SRE team consists of 50-60% Google software engineers. The other 40-50% are personnel who have software engineering skills but in addition, also have skills related to UNIX/Linux system internals and networking expertise. The team composition forced two behavioral patterns that propelled the team forward:

- Team members were quickly bored of performing tasks or responding to events manually.

- Team members had the capability to write software and provide an engineering solution to avoid repetitive manual work even if the solution is complicated.

In simple terminology, SRE practices evolved when a team of software engineers ran a service reliably in production and automated systems by using engineering practices. This raises some critical questions. How is SRE different from DevOps? Which is better? This will be covered in the upcoming sub-sections.

From Google's viewpoint, DevOps is a philosophy rather than a development methodology. It aims to close the gap between software development and software operations. DevOps' key pillars clarify what needs to be done to achieve collaboration, cohesiveness, flexibility, reliability, and consistency.

SRE's approach toward DevOps' key pillars

DevOps doesn't put forward a clear path or mechanism for how it needs to be done. Google's SRE approach is a concrete or prescriptive way to solve problems that the DevOps philosophy addresses. Google describes the relationship between SRE and DevOps using an analogy:

> *If you think of DevOps like an interface in a programming language, class SRE implements DevOps.*

(*Google Cloud, SRE vs. DevOps: competing standards or close friends?*, https://cloud.google.com/blog/products/gcp/sre-vs-devops-competing-standards-or-close-friends)

Let's look at how SRE implements DevOps and approaches the DevOps key pillars:

- **Reduces organizational silos**: SRE reduces organizational silos by sharing ownership between developers and operators. Both teams are involved in the product/service life cycle from the start. Together they define **Service-Level Objectives (SLOs)**, **Service-Level Indicators (SLIs)**, and **error budgets** and share the responsibility to determine the reliability, work priority, and release cadence of new features. This promotes a shared vision and improves communication and collaboration.

- **Accepts failure as normal**: SRE accepts failure as normal by conducting blameless postmortems, which includes detailed analysis without any reference to a person. **Blameless postmortems** help to understand the reasons for failure, identifying preventive actions, and ensuring that a failure for the same reason doesn't re-occur. The goal is to identify the root cause and process but not to focus on individuals. This helps to promote psychological safety. In most cases, failure is the result of a missing SLO or targets and incidents are tracked using specific indicators as a function of time or SLI.

- **Implements gradual change**: SRE implements gradual changes by limited **canary rollouts** and eventually reduces the cost of failures. Canary rollouts refer to the process of rolling out changes to a small percentage of users in production before making them generally available. This ensures that the impact is limited to a small set of users and gives us the opportunity to capture feedback on the new rollouts.

- **Leverages tooling and automation**: SRE leverages tooling and automation to reduce **toil** or the amount of manual repetitive work, and it eventually promotes speed and consistency. Automation is a force multiplier. However, this can create a lot of resistance to change. SRE recommends handling this resistance to change by understanding the psychology of change.

- **Measures everything**: SRE promotes data-driven decision making, encourages goal setting by measuring and monitoring critical factors tied to the health and reliability of the system. SRE also measures the amount of manual, repetitive work spent. Measuring everything is key for setting up SLOs and **Service-Level Agreements (SLAs)** and reducing toil.

This wraps up our introduction to SRE's approach to DevOps key pillars; we referred to jargon such as SLI, SLO, SLA, error budget, toil, and canary rollouts. These will be introduced in the next sub-section.

Introducing SRE's key concepts

SRE implements the DevOps philosophy via several key concepts, such as SLI, SLO, SLA, error budget, and toil.

Becoming familiar with SLI, SLO, and SLA

Before diving into the definitions of SRE terminology – specifically SLI, SLO, and SLA – this sub-section attempts to introduce this terminology through a relatable example.

Let's consider that you are a paid consumer for a video streaming service. As a paid consumer, you will have certain expectations from the service. A key aspect of that expectation is that the service needs to be available. This means when you try to access the website of the video streaming service via any permissible means, such as mobile device or desktop, the website needs to be accessible and the service should always work.

If you frequently encounter issues while accessing the service, either because the service is experiencing high traffic or the service provider is adding new features, or for any other reason, you will not be a happy consumer. Now, it is possible that some users can access this service at a moment in time but some users are unable to access it at the same moment in time. Those users who are able to access it are happy users and users who are unable to access it are sad users.

> **Availability**
> The first and most critical feature that a service should provide is **availability**. Service availability can also be referred to as its uptime. Availability is the ability of an application or service to run when needed. If a system is not running, then the system will fail.

Let's assume that you are a happy user. You can access the service. You can create a profile, browse titles, filter titles, watch reviews for specific titles, add videos to your watchlist, play videos, or add reviews to viewed videos. Each of these actions performed by you as a user can be categorized as a **user journey**. For each user journey, you will have certain expectations:

- If you try to browse titles under a specific category, say *comedy*, you would expect that the service loads the titles *without any delay*.

- If you select a title that you would like to watch, you would expect to watch the video *without any buffering*.

- If you would like to watch a livestream, you would expect the stream contents to be as *fresh* as possible.

Let's explore the first expectation. When you as a user tries to browse titles under *comedy*, how fast is fast enough?

Some users might expect to display the results within 1 second, and some might expect it in 200 ms and some others in 500 ms. So, the expectation needs to be quantifiable and for it to be quantifiable, it needs to be measurable. The expectation should be set to a value where most of the users will be happy. It should also be measured for a specific duration (say 5 minutes) and should be met over a period (say 30 days). It should not be a one-time event. If the expectation is not met over a period users expect, the service provider takes on some accountability and addresses the users' concerns either by issuing a refund or adding extra service credits.

For a service to be reliable, the service needs to have key characteristics based on expectations from user journeys. In this example, the key characteristics that the user expects are **latency**, **throughput**, and **freshness**.

Reliability

Reliability is the ability of an application or service to perform a specific function within a specific time without failures. If a system cannot perform its intended function, then the system will fail.

So, to summarize the example of a video streaming service, as a user you will expect the following:

- The service is available.

- The service is reliable.

Now, let's introduce SRE terminology with respect to the preceding example before going into their formal definitions:

- Expecting the service to be available or expecting the service to meet a specific amount of latency, throughput, or freshness, or any other characteristic that is critical to the user journey, is known as SLI.

- Expecting the service to be available or reliable for a certain target level over a specific period is SLO.

- Expecting the service to meet a pre-defined customer expectation, the failure of which results in a refund or credits, is SLA.

Let's move on from this general understanding of these concepts and explore how Google views them by introducing SRE's technical practices.

SRE's technical practices

SRE specifically prescribes the usage of specific technical tools or practices that will help to define, measure, and track service characteristics such as availability and reliability. These are referred to as SRE technical practices and specifically refer to SLIs, SLOs, SLAs, error budget, and toil. These are introduced in the following sections with significant insights.

Service-Level Indicator (SLI)

Google SRE has the following definition for SLI:

> *SLI is a carefully defined quantitative measure of some aspect of the level of service that is provided.*

(*Betsy Beyer, Chris Jones, Jennifer Petoff,* & *Niall Murphy, Site Reliability Engineering, O'REILLY*)

Most services consider latency or throughput as key aspects of a service based on related user journeys. SLI is a specific measurement of these aspects where raw data is aggregated or collected over a measurement window and represented as a rate, average, or percentile

Let's now look at the characteristics of SLIs:

- It is a direct measurement of a service performance or behavior.

- Refers to measurable metrics over time.

- Can be aggregated and turned to rate, average, or percentile.

- Used to determine the level of availability. SRE considers availability as the prerequisite to success.

SLI can be represented as a formula:

$$SLI = \frac{good\ events}{valid\ events} * 100$$

For systems serving requests over HTTPS, validity is often determined by request parameters such as hostname or requested path to scope the SLI to a particular set of serving tasks, or response handlers. For data processing systems, validity is usually determined by the selection of inputs to scope the SLI to a subset of data. Good events refer to the expectations from the service or system.

Let's look at some examples of SLIs:

- **Request latency**: The time taken to return a response for a request should be less than 100 ms.

- **Failure rate**: The ratio of unsuccessful requests to all received requests should be greater than 99%.

- **Availability**: Refers to the uptime check on whether a service is available or not at a particular point in time.

Service-Level Objective (SLO)

Google SRE uses the following definition for SLO:

> *Service level objectives (SLOs) specify a target level for the reliability of your service.*

(*Betsy Beyer, Chris Jones, Jennifer Petoff, & Niall Murphy, Site Reliability Engineering, O'REILLY*)

Customers have specific expectations from a service and these expectations are characterized by specific indicators or SLIs that are tailored per the user journey. SLOs are a way to measure customer happiness and their expectations by ensuring that the SLIs are consistently met and are potentially reported before the customer notices an issue.

Let's now look at the characteristics of SLOs:

- Identifies whether a service is reliable enough.

- Directly tied to SLIs. SLOs are in fact measured by using SLIs.

- Can either be a single target or a range of values for the collection of SLIs.

- If the SLI refers to metrics over time, which details the health of a service, then SLOs are agreed-upon bounds on how often the SLIs must be met.

Let's see how they are represented as a formula:

$SLI \leq target$ target OR $lower\ bound \leq SLI \leq upper\ bound$

SLO can best be represented either as a specific target value or as a range of values for an SLI for a specific aspect of a service, such as latency or throughput, representing the acceptable lower bound and possible upper bound that is valid over a specific period. Given that SLIs are used to measure SLOs, SLIs should be within the target or between the range of acceptable values

Let's look at some examples of SLOs:

- **Request latency**: 99.99% of all requests should be served under 100 ms over a period of 1 month or 99.9% of all requests should be served between 75 ms and 125 ms for a period of 1 month.

- **Failure rate**: 99.9% of all requests should have a failure rate of 99% over 1 year.
- **Availability**: The application should be usable for 99.95% of the time over 24 hours.

Service-Level Agreement (SLA)

Google SRE uses the following definition for SLAs:

> *SLA is an explicit or implicit contract with your users that includes consequences of meeting (or missing) the SLOs they contain.*

(Betsy Beyer, Chris Jones, Jennifer Petoff, & Niall Murphy, Site Reliability Engineering, O'REILLY)

An SLA is an external-facing agreement that is provided to the consumer of a service. The agreement clearly lays out the minimum expectations that the consumer can expect from the service and calls out the consequences that the service provider needs to face if found in violation. The consequences are generally applied in terms of refund or additional credits to the service consumer.

Let's now look at the characteristics of SLAs:

- SLAs are based on SLOs.
- Signifies the business factor that binds the customer and service provider.
- Represents the consequences of what happens when availability or customer expectation fails.
- Are more lenient than SLOs to trigger early alarms as these are the minimum expectations that the service should meet.

SLAs' priority in comparison to SLOs can be represented as follows:

$SLAs < SLOs$

Let's look at some examples of SLAs:

- **Latency**: 99% of all requests per day should be served under 150 ms; otherwise, 10% of the daily subscription fee will be refunded.
- **Availability**: The service should be available with an uptime commitment of 99.9% in a 30-day period; else 4 hours of extra credit will be added to the user account.

Error budgets

Google SRE defines error budgets as follows:

> *A quantitative measurement shared between the product and the SRE*
> *teams to balance innovation and stability.*

(*Betsy Beyer, Chris Jones, Jennifer Petoff,* & *Niall Murphy, Site Reliability Engineering,*
O'REILLY)

While a service needs to be reliable, it should also be mindful that if new features are not added to the service, then users might not continue to use it. A 100% reliable service will imply that the service will not have any downtime. This means that it will be increasingly difficult to add innovation via new features that could potentially attract new customers and lead to an increase in revenue. Getting to 100% reliability is expensive and complex. Instead, it's recommended to find the unique value for service reliability where customers feel that the service is reliable enough.

Unreliable systems can quickly erode users' confidence. So, it's critical to reduce the chance of system failure. SRE aims to balance the risk of unavailability with the goals of rapid innovation and efficient service operations so that users' overall happiness – with features, service, and performance – is optimized.

The error budget is basically the inverse of availability, and it tells us how unreliable our service is allowed to be. If your SLO says that 99.9% of requests should be successful in a given quarter, your error budget allows 0.1% of requests to fail. This unavailability can be generated because of bad pushes by the product teams, planned maintenance, hardware failures, and so on:

$Error\ Budget = 100\% - SLO$

Important note

The relationship between the error budget and actual allowed downtime for a service is as follows:

If SLO = 99.5%, then error budget = 0.5% = 0.005

Allowed downtime per month = 0.005 * 30 days/month * 24 hours/day * 60 minutes/hour = 216 minutes/month

---

The following table represents the allowed downtime for a specific time period to achieve a certain level of availability. For downtime information calculation for a specific availability level (other than the following mentioned), refer to `https://availability.sre.xyz/`:

Availability level	Downtime per year	Downtime per month	Downtime per day	Downtime per hour
90%	36.52 days	3.04 days	2.40 hours	6.00 minutes
99% (2 9's)	3.65 days	7.30 hours	14.40 minutes	36.00 seconds
99.5%	1.83 days	3.65 hours	7.20 minutes	18.00 seconds
99.9% (3 9's)	8.77 hours	43.83 minutes	1.44 minutes	3.60 seconds
99.99% (4 9's)	52.59 minutes	4.38 minutes	8.64 seconds	0.36 seconds
99.999% (5 9's)	5.26 minutes	26.30 seconds	0.86 seconds	0.04 seconds

There are advantages to defining the error budget:

- Release new features while keeping an eye on system reliability.
- Roll out infrastructure updates.
- Plan for inevitable failures in networks and other similar events.

Despite planning error budgets, there are times when a system can overshoot it. In such cases, there are a few things that occur:

- Release of new features is temporarily halted.
- Increased focus on dev, system, and performance testing.

Toil

Google SRE defines toil as follows:

> *Toil is the kind of work tied to running a production service that tends to be manual, repetitive, automatable, tactical, devoid of enduring value and that scales linearly as a service grows.*

(*Betsy Beyer, Chris Jones, Jennifer Petoff, & Niall Murphy, Site Reliability Engineering, O'REILLY*)

Here are the characteristics of toil:

- **Manual**: Act of manually initiating a script that automates a task.
- **Repetitive**: Tasks that are repeated multiple times.

- **Automatable**: Human executing a task instead of a machine, especially if a machine can execute with the same effectiveness.

- **Tactical**: Reactive tasks originating out of an interruption (such as pager alerts), rather than strategy-driven proactive tasks, are considered toil.

- **No enduring value**: Tasks that do not change the effective state of the service after execution.

- **Linear growth**: Tasks that grow linearly with an increase in traffic or service demand.

Toil is generally confused with **overhead**. Overhead is not the same as toil. Overhead is referred to as administrative work that is not tied to running a service, but toil refers to repetitive work that can be reduced by automation. Automation helps to lower burnout, increase team morale, increase engineering standards, improve technical skills, standardize processes, and reduce human error. Examples of tasks that represent overhead and not toil are email, commuting, expense reports, and meetings.

Canary rollouts

SRE prescribes implementing gradual change by using canary rollouts, where the concept is to introduce a change to a small portion of users to detect any imminent danger.

To elaborate, when there is a large service that needs to be sustained, it's preferable to employ a production change with unknown impact to a small portion to identify any potential issue. If any issues are found, the change can be reversed, and the impact or cost is much less than if the change was rolled out to the whole service.

The following two factors should be considered when selecting the canary population:

- The size of the canary population should be small enough that it can be quickly rolled back in case an issue arises.

- The size of the canary population should be large enough that it is a representative subset of the total population.

This concludes a high-level overview of important SRE technical practices. The next section details SRE cultural practices that are key to embrace SRE across an organization and are also critical to efficiently handle change management.

SRE's cultural practices

Defining SLIs, SLOs, and SLAs for a service, using error budgets to balance velocity (the rate at which changes are delivered to production) and reliability, identifying toil, and using automation to eliminate toil forms SRE's technical practices. In addition to these technical practices, it is important to understand and build certain cultural practices that eventually support the technical practices. Cultural practices are equally important to reduce silos within IT teams, as they can reduce the incompatible practices used by individuals within the team. The first cultural practice that will be discussed is the need for a unifying vision.

Need for a unifying vision

Every company needs a vision and a team's vision needs to align with the company's vision. The company's vision is a combination of core values, the purpose, the mission, strategies, and goals:

- **Core values**: Values refer to a team member's commitment to personal/ organizational goals. It also reflects on how members operate within a team by building trust and psychological safety. This creates a culture where the team is open to learning and willing to take risks.

- **Purpose**: A team's purpose refers to the core reason that the team exists in the first place. Every team should have a purpose in the larger context of the organization.

- **Mission**: A team's mission refers to a well-articulated, clear, compelling, and unified goal.

- **Strategy**: A team's strategy refers to a plan on how the team will realize its mission.

- **Goals**: A team's goal gives more detailed and specific insights into what the team wants to achieve. Google recommends the use of **Objectives and Key Results (OKRs)**, which are a popular goal-setting tool in large companies.

Once a vision statement is established for the company and the team, the next cultural practice is to ensure there is efficient collaboration and communication within the team and across cross-functional teams. This will be discussed next.

Collaboration and communication

Communication and collaboration are critical given the complexity of services and the need for these services to be globally accessible. This also means that SRE teams should be globally distributed to support services in an effective manner. Here are some SRE prescriptive guidelines:

- **Service-oriented meetings**: SRE teams frequently review the state of the service and identify opportunities to improve and increase awareness among stakeholders. The meetings are mandatory for team members and typically last 30-60 minutes, with a defined agenda such as discussing recent paging events, outages, any required configuration changes.

- **Balanced team composition**: SRE teams are spread across multiple countries and multiple time zones. This enables them to support a globally available system or service. The SRE team composition typically includes a technical lead (to provide technical guidance), a manager (who runs performance management), and a project manager, who collaborate across time zones.

- **Involvement throughout the service life cycle**: SRE teams are actively involved throughout the service life cycle across various stages such as *architecture and design, active development, limited availability, general availability*, and *depreciation*.

- **Establish rules of engagement**: SRE teams should clearly describe what channels should be used for what purpose and in what situations. This brings in a sense of clarity. SRE teams should use a common set of tools for creating and maintaining artifacts

- **Encourage blameless postmortem**: SRE encourages a blameless postmortem culture, where the theme is to learn from failure and the focus is on identifying the root cause of the issue rather than on individuals. A well-written postmortem report can act as an effective tool for driving positive organizational changes since the suggestions or improvements mentioned in the report can help to tune up existing processes

- **Knowledge sharing**: SRE teams prescribe knowledge sharing through specific means such as encouraging cross-training, creation of a volunteer teaching network, and sharing postmortems of incidents in a way that fosters collaboration and knowledge sharing.

The preceding guidelines, such as knowledge sharing along with the goal to reduce paging events or outages by creating a common set of tools, increase resistance among individuals and team members. This might also create a sense of insecurity. The next cultural practice elaborates on how to encourage psychological safety and reduce resistance to change.

Encouraging psychological safety and reducing resistance to change

SRE prescribes automation as an essential cornerstone to apply engineering principles and reduce manual work such as toil. Though eliminating toil through automation is a technical practice, there will be huge resistance to performing automation. Some may resist automation more than others. Individuals may feel as though their jobs are in jeopardy, or they may disagree that certain tasks need not be automated. SRE prescribes a cultural practice to reduce the resistance to change by building a psychologically safe environment.

In order to build a psychologically safe environment, it is first important to communicate the importance of a specific change. For example, if the change is to automate this year's job away, here are some reasons on how automation can add value:

- Provides consistency.

- Provides a platform that can be extended and applied to more systems.

- Common faults can be easily identified and resolved more quickly.

- Reduces cost by identifying problems as early in the life cycle as possible, rather than finding them in production.

Once the reason for the change is clearly communicated, here are some additional pointers that will help to build a *psychologically safe environment*:

- Involve team members in the change. Understand their concerns and empathize as needed.

- Encourage critics to openly express their fears as this adds a sense of freedom to team members to freely express their opinions.

- Set realistic expectations.

- Allow team members to adapt to new changes.

- Provide them with effective training opportunities and ensure that training is engaging and rewarding.

This completes an introduction to key SRE cultural practices that are critical to implementing SRE's technical practices. Subsequently, this also completes the section on SRE where we introduced SRE, discussed its evolution, and elaborated on how SRE is a prescriptive way to practically implement DevOps key pillars. The next section discusses how to implement DevOps using Google Cloud services.

Cloud-native approach to implementing DevOps using Google Cloud

This section elaborates on how to implement DevOps using Google Cloud services with a focus on a cloud-native approach – an approach that uses cloud computing at its core to build highly available, scalable, and resilient applications.

Focus on microservices

A **monolith application** has a tightly coupled architecture and implements all possible features in a single code base along with the database. Though monolith applications can be designed with modular components, the components are still packaged at deployment time and deployed together as a single unit. From a CI/CD standpoint, this will potentially result in a single build pipeline. Fixing an issue or adding a new feature is an extremely time-consuming process since the impact is on the entire application. This decreases the release velocity and essentially is a nightmare for production support teams dealing with service disruption.

In contrast, a **microservice application** is based on service-oriented architecture. A microservice application divides a large program into several smaller, independent services. This allows the components to be managed by smaller teams as the components are more isolated in nature. The teams, as well as the service, can be independently scaled. Microservices fundamentally support the concept of incremental code change. With microservices, the individual components are deployable. Given that microservices are feature-specific, in the event of an issue, fault detection and isolation are much easier and hence service disruptions can be handled quickly and efficiently. This also makes it much more suitable for CI/CD processes and works well with the theme of *building reliable software faster*!

> **Exam tip**
>
> Google Cloud provides several compute services that facilitate the deployment of microservices as containers. These include App Engine flexible environment, Cloud Run, **Google Compute Engine (GCE)**, and **Google Kubernetes Engine (GKE)**. From a Google Cloud DevOps exam perspective, the common theme is to build containers and deploy containers using GKE. GKE will be a major focus area and will be discussed in detail in the upcoming chapters.

Cloud-native development

Google promotes and recommends application development using the following cloud-native principles:

- **Use microservice architectural patterns**: As discussed in the previous sub-section, the essence is to build smaller independent services that could be managed separately and be scaled granularly.

- **Treat everything as code**: This principle makes it easier to track, roll back code if required, and see the version of change. This includes source code, test code, automation code, and infrastructure as code.

- **Build everything as containers**: A container image can include software dependencies needed by the application, specific language runtimes, and other software libraries. Containers can be run anywhere, making it easier to develop and deploy. This allows developers to focus on code and ops teams will spend less time debugging and diagnosing differences in environments.

- **Design for automation**: Automated processes can repair, scale, and deploy systems faster than humans. As a critical first step, a comprehensive CI/CD pipeline is required that can automate the build, testing, and deployment process. In addition, the services that are deployed as containers should be configured to scale up or down based on outstanding traffic. Real-time monitoring and logging should be used as a source for automation since they provide insights into potential issues that could be mitigated by building proactive actions. The idea of automation can also be extended to automate the entire infrastructure using techniques such as **Infrastructure as Code (IaC)**.

- **Design components to be stateless wherever possible**: Stateless components are easy to scale up or down, repair a failed instance by graceful termination and potential replacement, roll back to an older instance in case of issues, and make load balancing a lot simpler since any instance can handle any request. Any need to store persistent data should happen outside the container, such as storing files using Cloud Storage, storing user sessions through Redis or Memcached, or using persistent disks for block-level storage.

Google Cloud provides two approaches for cloud-native development – **serverless** and **Kubernetes**. The choice comes down to focus on infrastructure versus business logic:

- **Serverless** (via Cloud Run, Cloud Functions, or App Engine): Allows us to focus on the business logic of the application by providing a higher level of abstraction from an infrastructure standpoint.

- **Kubernetes** (via GKE): Provides higher granularity and control on how multiple microservices can be deployed, how services can communicate with each other, and how external clients can interact with these services.

Managed versus serverless service

Managed services allow operations related to updates, networking, patching, high availability, automated backups, and redundancy to be managed by the cloud provider. Managed services are not serverless as it is required to specify a machine size and the service mandates to have a minimal number of VMs/ nodes. For example, it is required to define the machine size while creating a cloud SQL instance, but updates and patches can be configured to be managed by Google Cloud.

Serverless services are **managed** but do not require reserving a server upfront or keeping it running. The focus is on the business logic of the application with the possibility of running or executing code only when needed. Examples are *Cloud Run*, *Cloud Storage*, *Cloud Firestore*, and *Cloud Datastore*.

Continuous integration in GCP

Continuous integration forms the **CI** of the CI/CD process and at its heart is the culture of submitting smaller units of change frequently. Smaller changes minimize the risk, help to resolve issues quickly, increase development velocity, and provide frequent feedback. The following are the building blocks that make up the CI process:

- **Make code changes**: By using the IDE of choice and possible cloud-native plugins
- **Manage source code**: By using a single shared code repository
- **Build and create artifacts**: By using an automated build process
- **Store artifacts**: By storing artifacts such as container images in a repository for a future deployment process

Google Cloud has an appropriate service for each of the building blocks that allows us to build a GCP-native CI pipeline (refer to *Figure 1.2*). The following is a summary of these services, which will be discussed in detail in upcoming chapters:

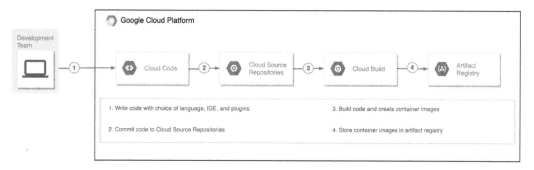

Figure 1.2 – CI in GCP

Let's look at these stages in detail.

Cloud Code

This is the GCP service to write, debug, and deploy cloud-native applications. Cloud Code provides extensions to IDEs such as *Visual Studio Code* and the *JetBrains suite* of IDEs that allows to rapidly iterate, debug, and run code on Kubernetes and Cloud Run. Key features include the following:

- Speed up development and simplify local development
- Extend to production deployments on GKE or Cloud Run and allow debugging deployed applications
- Deep integration with Cloud Source Repositories and Cloud Build
- Easy to add and configure Google Cloud APIs from built-in library manager

Cloud Source Repositories

This is the GCP service to manage source code. It provides Git version control to support the collaborative development of any application or service. Key features include the following:

- Fully managed private Git repository
- Provides one-way sync with Bitbucket and GitHub source repositories
- Integration with GCP services such as Cloud Build and Cloud Operations
- Includes universal code search within and across repositories

Cloud Build

This is the GCP service to build and create artifacts based on commits made to source code repositories such as GitHub, Bitbucket, or Google's Cloud Source Repositories. These artifacts can be container or non-container artifacts. The GCP DevOps exam's primary focus will be on container artifacts. Key features include the following:

- Fully serverless platform with no need to pre-provision servers or pay in advance for additional capacity. Will scale up and down based on load

- Includes Google and community builder images with support for multiple languages and tools

- Includes custom build steps and pre-created extensions to third-party apps that enterprises can easily integrate into their build process

- Focus on security with vulnerability scanning and the ability to define policies that can block the deployment of vulnerable images

Container/Artifact Registry

This is the GCP construct to store artifacts that include both container (Docker images) and non-container artifacts (such as Java and Node.js packages). Key features include the following:

- Seamless integration with Cloud Source Repositories and Cloud Build to upload artifacts to Container/Artifact Registry.

- Ability to set up a secure private build artifact storage on Google Cloud with granular access control.

- Create multiple regional repositories within a single Google Cloud project.

Continuous delivery/deployment in GCP

Continuous delivery/deployment forms the **CD** of the CI/CD process and at its heart is the culture of continuously delivering production-ready code or deploying code to production. This allows us to release software at high velocity without sacrificing quality.

GCP offers multiple services to deploy code, such as *Compute Engine*, *App Engine*, *Kubernetes Engine*, *Cloud Functions*, and *Cloud Run*. The focus of this book will be on GKE and Cloud Run. This is in alignment with the Google Cloud DevOps exam objectives.

The following figure summarizes the different stages of continuous delivery/deployment from the viewpoint of appropriate GCP services:

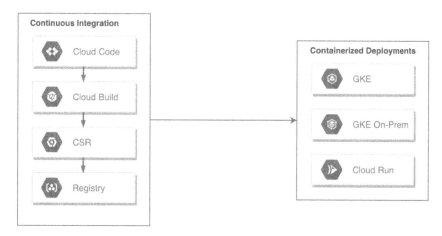

Figure 1.3 – Continuous delivery/deployment in GCP

Let's look at the two container-based deployments in detail.

Google Kubernetes Engine (GKE)

This is the GCP service to deploy containers. GKE is Google Cloud's implementation of the **CNCF Kubernetes project**. It's a managed environment for deploying, managing, and scaling containerized applications using Google's infrastructure. Key features include the following:

- Automatically provisions and manages a cluster's master-related infrastructure and abstracts away the need for a separate master node
- Automatic scaling of a cluster's node instance count
- Automatic upgrades of a cluster's node software
- Node auto-repair to maintain the node's health
- Native integration with Google's Cloud Operations for logging and monitoring

Cloud Run

This is a GCP-managed serverless platform that can deploy and run Docker containers. These containers can be deployed in either Google-managed Kubernetes clusters or on-premises workloads using *Cloud Run for Anthos*. Key features include the following:

- Abstracts away infrastructure management by automatically scaling up and down
- Only charges for exact resources consumed

- Native GCP integration with Google Cloud services such as Cloud Code, Cloud Source Repositories, Cloud Build, and Artifact Registry

- Supports event-based invocation via web requests with Google Cloud services such as Cloud Scheduler, Cloud Tasks, and Cloud Pub/Sub

Continuous monitoring/operations on GCP

Continuous Monitoring/Operations forms the feedback loop of the CI/CD process and at its heart is the culture of continuously monitoring or observing the performance of the service/application.

GCP offers a suite of services that provide different aspects of Continuous Monitoring/ Operations, aptly named **Cloud Operations** (formerly known as **Stackdriver**). Cloud Operations includes **Cloud Monitoring**, **Cloud Logging**, **Error Reporting**, and **Application Performance Management** (**APM**). APM further includes **Cloud Debugger**, **Cloud Trace**, and **Cloud Profiler**. Refer to the following diagram:

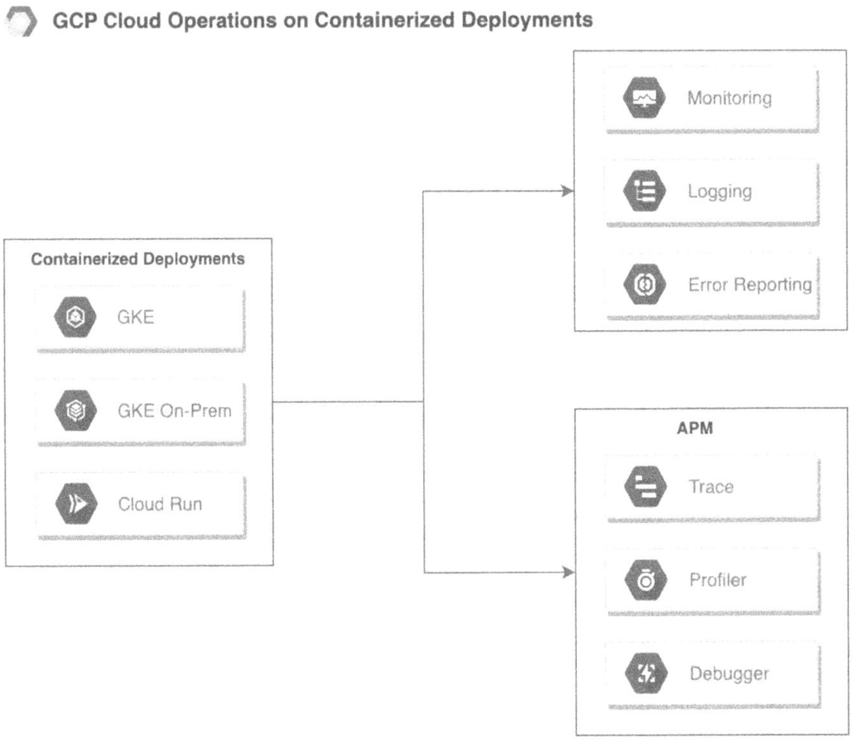

Figure 1.4 – Continuous monitoring/operations

Let's look at these operations- and monitoring-specific services in detail.

Cloud Monitoring

This is the GCP service that collects metrics, events, and metadata from Google Cloud and other providers. Key features include the following:

- Provides out-of-the-box default dashboards for many GCP services
- Supports uptime monitoring and alerting to various types of channels
- Provides easy navigation to drill down from alerts to dashboards to logs and traces to quickly identify the root cause
- Supports non-GCP environments with the use of agents

Cloud Logging

This is the GCP service that allows us to store, search, analyze, monitor, and alert on logging data and events from **Google Cloud** and **Amazon Web Services**. Key features include the following:

- A fully managed service that performs at scale with sub-second ingestion latency at terabytes per second
- Analyzes log data across multi-cloud environment from a single place
- Ability to ingest application and system log data from thousands of VMs
- Ability to create metrics from streaming logs and analyze log data in real time using BigQuery

Error Reporting

This is the GCP service that aggregates, counts, analyzes, and displays application errors produced from running cloud services. Key features include the following:

- Dedicated view of error details that include a time chart, occurrences, affected user count, first and last seen dates, and cleaned exception stack trace
- Lists out the top or new errors in a clear dashboard
- Constantly analyzes exceptions and aggregates them into meaningful groups
- Can translate the occurrence of an uncommon error into an alert for immediate attention

Application Performance Management

This is the GCP service that combines monitoring and troubleshooting capabilities of Cloud Logging and Cloud Monitoring with Cloud Trace, Cloud Debugger, and Cloud Profiler, to help reduce latency and cost and enable us to run applications more efficiently. Key features include the following:

- A distributed tracing system (via Cloud Trace) that collects latency data from your applications to identify performance bottleneck

- Inspects a production application by taking a snapshot of the application state in real time, without stopping or slowing down (via Cloud Debugger), and provides the ability to inject log messages as part of debugging

- Low-impact production profiling (via Cloud Profiler) using statistical techniques, to present the call hierarchy and resource consumption of relevant function in an interactive flame graph

Bringing it all together – building blocks for a CI/CD pipeline in GCP

The following figure represents the building blocks that are required to build a CI/CD pipeline in GCP:

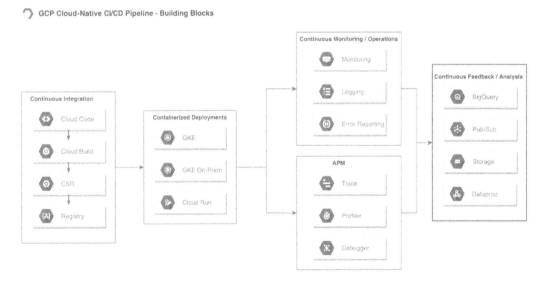

Figure 1.5 – GCP building blocks representing the DevOps life cycle

In the preceding figure, the section for **Continuous Feedback/Analysis** represents the GCP services that are used to analyze or store information obtained during Continuous Monitoring/Operations either from an event-driven or compliance perspective. This completes the section on an overview of Google Cloud services that can be used to implement the key stages of the DevOps life cycle using a cloud-native approach with emphasis on decomposing a complex system into microservices that can be independently tested and deployed.

Summary

In this chapter, we learned about DevOps practices that break down the metaphorical wall between developers (who constantly want to push features to production) and operators (who want to run the service reliably).

We learned about the DevOps life cycle, key pillars of DevOps, how Google Cloud implements DevOps through SRE, and Google's cloud-native approach to implementing DevOps. We learned about SRE's technical and cultural practices and were introduced to key GCP services that help to build the CI/CD pipeline. In the next chapter, we will take an in-depth look at SRE's technical practices such as SLI, SLO, SLA, and error budget.

Points to remember

The following are some important points:

- If DevOps is a philosophy, SRE is a prescriptive way of achieving that philosophy: *class SRE* implements DevOps.

- SRE balances the velocity of development features with the risk to reliability.

- SLA represents an external agreement and will result in consequences when violated.

- SLOs are a way to measure customer happiness and their expectations.

- SLIs are best expressed as a proportion of all successful events to valid events.

- Error budget is the inverse of availability and depicts how unreliable a service is allowed to be.

- Toil is manual work tied to a production system but is not the same as overhead.

- The need for unifying vision, communication, and collaboration with an emphasis on blameless postmortems and the need to encourage psychological safety and reduce resistance to change are key SRE cultural practices.

- Google emphasizes the use of microservices and cloud-native development for application development.

- Serverless services are managed but managed services are necessarily not serverless.

Further reading

For more information on GCP's approach toward DevOps, read the following articles:

- DevOps: `https://cloud.google.com/devops`

- SRE: `https://landing.google.com/sre/`

- CI/CD on Google Cloud: `https://cloud.google.com/docs/ci-cd`

Practice test

Answer the following questions:

1. Which of the following represents a sequence of tasks that is central to user experience and is crucial to service?

    a) User story

    b) User journey

    c) Toil

    d) Overhead

2. If the SLO for the uptime of a service is set to 99.95%, what is the possible SLA target?

    a) 99.99

    b) 99.95

    c) 99.96

    d) 99.90

3. Which of the following accurately describes the equation for SLI?

    a) Good events / Total events

    b) Good events / Total events * 100

    c) Good events / Valid events

    d) Good events / Valid events * 100

4. Which of the following represents a carefully defined quantitative measure of some aspect of the level of service?

   a) SLO

   b) SLI

   c) SLA

   d) Error budget

5. Select the option used to calculate the error budget.

   a) (100 – SLO) * 100

   b) 100 – SLI

   c) 100 – SLO

   d) (100 – SLI) * 100

6. Which set of Google services accurately depicts the continuous feedback loop?

   a) Monitoring, Logging, Reporting

   b) Bigtable, Cloud Storage, BigQuery

   c) Monitoring, Logging, Tracing

   d) BigQuery, Pub-Sub, Cloud Storage

7. In which of the following "continuous" processes are changes automatically deployed to production without manual intervention?

   a) Delivery

   b) Deployment

   c) Integration

   d) Monitoring

8. Select the option that ranks the compute services from a service that requires the most management needs with the highest customizability to a service with fewer management needs and the lowest customizability.

   a) Compute Engine, App Engine, GKE, Cloud Functions

   b) Compute Engine, GKE, App Engine, Cloud Functions

   c) Compute Engine, App Engine, Cloud Functions, GKE

   d) Compute Engine, GKE, Cloud Functions, App Engine

9. Awesome Incorporated is planning to move their on-premises CI pipeline to the cloud. Which of the following services provides a private Git repository hosted on GCP?

    a) Cloud Source Repositories

    b) Cloud GitHub

    c) Cloud Bitbucket

    d) Cloud Build

10. Your goal is to adopt SRE cultural practices in your organization. Select two options that could help to achieve this goal.

    a) Launch and iterate.

    b) Enable daily culture meetings.

    c) Ad hoc team composition.

    d) Create and communicate a clear message.

Answers

1. (b) – User journey

2. (d) – 99.90

3. (d) - Good events / Valid events * 100

4. (b) - SLI

5. (c) – 100 – SLO

6. (d) – BigQuery, Pub-Sub, Cloud Storage

7. (b) – Deployment (forming continuous deployment)

8. (b) – Compute Engine, GKE, App Engine, Cloud Functions

9. (a) – Cloud Source Repositories

10. (a) and (d) – Launch and iterate. Create and communicate a clear message.

2
SRE Technical Practices – Deep Dive

Reliability is the most critical feature of a service or a system and should be aligned with business objectives. This alignment should be tracked constantly, meaning that the alignment needs measurement. **Site reliability engineering** (**SRE**) prescribes specific technical tools or practices that will help in measuring characteristics that define and track reliability. These tools are **service-level agreements** (**SLAs**), **service-level objectives** (**SLOs**), **service-level indicators** (**SLIs**), and **error budgets**.

SLAs represent an external agreement with customers about the reliability of a service. SLAs should have consequences if violated (that is, the service doesn't meet the reliability expectations), and the consequences are often monetary in nature. To ensure SLAs are never violated, it is important to set thresholds. Setting these thresholds ensures that an incident is caught and potentially addressed before repeated occurrences of similar or the same events breach the SLA. These thresholds are referred to as SLOs.

SLOs are specific numerical targets to define reliability of a system, and SLOs are measured using SLIs. SLIs are a quantitative measure of the level of service provided over a period. Error budgets are calculated based on SLOs (that are based on SLIs) and essentially are the inverse of availability, representing a quantifiable target as to how much a service can be unreliable. All these tools or technical practices need to work in tandem, and each one is dependent on the other. SRE uses these technical practices to maintain the balance between innovation and system reliability and thus achieve the eventual goal—*build reliable software faster*.

Chapter 1, *DevOps, SRE, and Google Cloud Services for CI/CD*, introduced SRE technical practices—SLAs, SLOs, SLIs, and error budgets. This chapter will deep dive into these technical practices. In this chapter, we're going to cover the following main topics:

- Defining SLAs
- Defining reliability expectations via SLOs
- Exploring SLIs
- Understanding error budgets
- Eliminating toil through automation
- Illustrating the impact of SLAs, SLOs, and error budgets relative to SLI

Defining SLAs

An SLA is a promise made to a user of a service to indicate that the availability and reliability of the service should meet a certain level of expectation. An SLA details a certain level of performance or expectation from the service.

Key jargon

There are certain components that go into defining which agreements can be considered as an SLA. These are referred to with specific jargon and are elaborated, as mentioned, in the following sections.

Service provider and service consumer

The party that represents the service provider and service consumer can differ based on the context and nature of the service. For a consumer-facing service such as video streaming or web browsing, a service consumer refers to the end user consuming the service and a service provider refers to the organization providing the service. On the other hand, for an enterprise-grade service such as a **human resource** (**HR**) planning system, a service consumer refers to the organization consuming the service and a service provider refers to the organization providing the service.

Service performance or expectations

An organization or end user consuming a service will have certain expectations in terms of service behavior, such as **availability** (or **uptime**), **responsiveness**, **durability**, and **throughput**.

Agreement – implicit or explicit

An agreement or contract can be either implicit or explicit in nature. An example of an implicit contract is a non-commercial service such as Google Search. Google has a goal to provide a fluid search experience to all its users but hasn't signed an explicit agreement with the end user. If Google misses its goal, then users will not have a good experience. A repeat of such incidents will impact Google's reputation, as users might prefer to use an alternate search engine.

An example of an explicit contract is a commercial service such as *Netflix* or a paid enterprise-grade service such as *Workday*. In such scenarios, legal agreements are written that include consequences in case the service expectations are not met. Common consequences include financial implications or service credits.

This concludes an introduction to key jargon with respect to SLAs. The next subsection elaborates on the blueprint for a well-defined SLA.

Blueprint for a well-defined SLA

Having a well-defined SLA is critical to its success. Here are some factors that could be used as a blueprint for a well-defined SLA:

- **Involve the right parties**: SLAs are usually written by people who are not directly tied to the implementation of the service and hence might result in promises that are difficult to measure. *SLAs should be set between business and product owners.* However, SRE recommends that before SLAs are set, product owners should work with development and SRE teams to identify the expectation threshold that can be delivered by the service. This ensures that product owners work closer with the implementation teams and know what's acceptable and what's not realistic from a service viewpoint.

- **Expectations need to be measurable**: Service expectations such as availability or reliability characteristics in terms of stability, responsiveness, and durability should be quantifiable and measurable. Service expectations should be monitored by configuring monitoring systems and tracking specific metrics, and alerting should be configured to trigger alerts in case the expectations are violated.

- **Avoid ambiguity**: The jargon used while defining SLAs can sometimes be ambiguous. For example, consider an SLA that promises a client-initiated incident to be resolved within X hours from the time it's reported. If the client or customer either provided the details long after the incident was first reported or never provided the details at all, then it is possible that the service provider will not be able to resolve the incident. In this situation, the SLA should clearly state the stipulations that qualify for not meeting the SLA, and such scenarios should be excluded. This provides a clearer approach.

SLIs drive SLOs, which inform SLAs

SLAs should focus on the minimum level of objectives a service should meet to keep customers happy. However, SLAs are strictly external targets and should not be used as internal targets by the implementation teams.

To ensure that SLAs are not violated, implementation teams should have target objectives that reflect user's expectations from the service. The target objectives from implementation teams are used as internal targets, and these are generally stricter than the external targets that were potentially set by product teams. The internal targets are referred to as SLOs and are used as a prioritization signal to balance release velocity and system reliability. These internal targets need to be specifically measured and quantified at a given point of time. The measurement should be done using specific indicators that reflects users' expectations, and such indicators are referred to as SLIs.

To summarize, for a service to perform reliably, the following criteria needs to be met:

- A specific condition should be met—represented by an SLI.
- The condition should be met for a specific period within a specific target range—represented by an SLO.
- If met, customers are happy, or else there will be consequences—represented by an SLA.

Let's look at a hypothetical example. Consider a requirement where a user's request/response time falls within a minimum time period. A latency metric can be used to represent the user's expectation. A sample SLA in this scenario can state that every customer will get a response within 1,000 **milliseconds** (**ms**). In this case, the SLO for this SLA must be stricter and can be set at 800 ms.

This completes the section on SLAs. We looked at the key constructs of an SLA, factors that could impact a well-defined SLA, and its impact on setting internal target objectives or SLOs, using specific indicators or SLIs that impact customer satisfaction. The next section transitions from an SLA to an SLO and its respective details.

Defining reliability expectations via SLOs

Service consumers (users) need a service to be reliable, and the reliability of the service can be captured by multiple characteristics such as availability, latency, freshness, throughput, coverage, and so on. From a user's perspective, a service is reliable if it meets their expectations. A critical goal of SRE is to measure everything in a quantitative manner. So, to measure, there is a need to represent user expectations quantitatively.

SRE recommends a specific technical practice called a SLO to specify a target level (numerical) to represent these expectations. Each service consumer can have a different expectation. These expectations should be measurable, and for that they should be quantifiable over a period. SLOs help to define a consistent level of user expectations where the measured user expectation should be either within the target level or should be within a range of values. In addition, SLOs are referred to as internal agreements and are often stricter than SLAs promised to the end users. This ensures that any potential issues are resolved before their repetitive occurrence results in violating the SLA.

SLOs are key to driving business decisions by providing a quantifiable way to balance release cadence of service features versus service reliability. This emphasis will be covered in the upcoming subsection.

SLOs drive business decisions

The need for revenue growth puts businesses under constant pressure to add new features and attract new users to their service. So, product managers usually dictate the requirement of these new features to development teams. Development teams build these requirements and hand them over to the operations team to stabilize. Development teams continue their focus on adding new features to a service rather than stabilizing existing ones. Operations teams tend to get overloaded since they are constantly firefighting to maintain the reliability of the existing service, in addition to rolling out new features. So, the most important question is: *If reliability is a feature of a system, then how can you balance reliability along with the release of other features?*

SLOs are the answer to how to maintain a balance between reliability and release velocity. SLOs allow us to define target levels for a reliable service. These target levels should be decided by all the stakeholders across an organization, including engineering teams (development and operations) and the product team. The agreed-upon target levels should reflect users' experiences while using the service. This allows monitoring systems to identify existing problems before users register complaints. SLOs should be treated more as a prioritization signal rather than an operational concern.

SLOs should be used as a primary driver for decision making. SLOs represent a common language for all reliability conversations that is based on actual metrics. This will allow a business to decide when to release new features versus when to continue their focus on the reliability of an existing service. It will also allow operations teams to have a streamlined set of goals, preventing ad-hoc actions to run the service, and eventually avoiding **operational overload**.

Operational overload is a term that describes the ongoing maintenance tasks that keep systems and services running at optimal performance. If a team is constantly interrupted by operations load and cannot make progress toward their key priorities, then the team is in a state of operational overload.

The main reason for a team to be in a state of operational overload is a lack of consensus on the level of reliability a service should support. This lack of consensus is apparent from development teams' focus on adding new features to a service rather than stabilizing existing ones.

SLOs must have strong backing from the executive team. In the case of missed SLO targets, there should be well-documented consequences that prioritize engineering efforts toward stabilizing the reliability of a service rather than working or releasing new features. SLOs are key to removing organization silos and create a sense of shared responsibility and ownership. SLOs drive incentives that organically invoke a thought process whereby developers start to care about service reliability and operators start to care about pushing new features out as quickly as possible. The recommended guidelines to set SLOs will be detailed in the upcoming subsection.

Setting SLOs – the guidelines

The journey or process to identify the right SLOs for a service is very complex. There are multiple aspects or guidelines that need to be considered. Each of these aspects is critical to set or define an SLO for a service.

The happiness test

SLO targets are always driven by quantifiable and measurable user expectations called SLIs. **The happiness test** is a good starting point to set SLO targets for a service. As per the test, the service should have target SLOs that barely meet the availability and reliability expectations of the users, as the following applies:

- If the service meets the target SLOs, then users are happy.

- If the service misses the target SLOs, then users are sad.

A target SLO for an average response time is defined as a range between 600 and 800 ms. If the average response time is less than 800 ms, then the service meets the target SLO, and users are happy. If the average response time is greater than 800 ms (even though it is less than stipulated in the SLA), then the service misses the target SLO and the users are sad. The following diagram illustrates an example where an SLA with respect to average response time for a request is set to 1,000 ms:

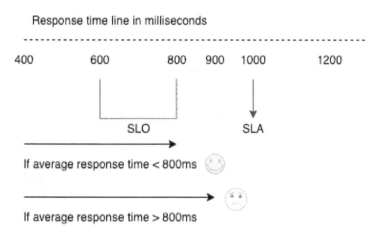

Figure 2.1 – Happy versus sad users based on target SLOs

100% reliability is the wrong target

Reliability is the most important feature of a service and reflects user happiness. However, setting 100% as the SLO or reliability target is not a realistic and reasonable goal for the following reasons:

- **Unable to improve or add new features to the service**: Maintaining customer happiness is tricky. Customers always look forward to new feature sets but also expect that the stability of the existing service will not be impacted. Adding new features to a running service can have the potential to introduce some amount of risk or unreliability. If SLO targets for a service are set to 100%, this implies that the service is always reliable, resulting in zero downtime. As a result, the service cannot tolerate any risk in terms of downtime, and inherently new features cannot be added to the service. If new features are not added to the service, users will be unhappy and will move to competitors that offers similar services with more feature sets.

- **Technologically unfeasible**: Running a service includes multiple components and dependencies. Some of these are internal, while some are external to the service. Though these components can be made redundant to achieve high availability, the dependencies result in complexities that would result in potential downtime. In addition, external components impact the availability of a service—for example, a mobile user cannot access a service if the mobile network provider has a dead zone at that specific location.

- **Exponentially expensive**: For every additional nine of reliability, the cost increases by 10 times. It's expensive to make a reliable system even more reliable. Being reliable enough is the wiser option.

Understanding reliability targets and their implications

As 100% is the wrong reliability target, it is important to find the optimal reliability target where the *service is reliable enough* for the user and there is an opportunity to update or add new features to the service.

Another perspective with which to look at reliability targets for a service is the amount of unreliability the service is willing to tolerate. Unreliability of the service is also referred to as the **downtime**.

Let's consider some reliability targets, as follows:

- A reliability target of 99.9% (also known as three nines of reliability) over a 30-day period will result in a maximum possible downtime of 42 minutes. This is enough time for a monitoring system to detect the issue, and is also enough time for a human to get involved and probably mitigate or resolve the issue.

- A reliability target of 99.99% (also known as four nines of reliability) over a 30-day period will result in a maximum possible downtime of 4.2 minutes. This is enough time for a monitoring system to detect the issue but is not enough time for a human to get involved, but probably enough time for a system to self-heal a complete outage.

- A reliability target of 99.999% (also known as five nines of reliability) over a 30-day period will result in a maximum possible downtime of 24 seconds. This extremely short duration is not enough to detect an issue or even attempt to self-heal.

The following table summarizes the possibility of detecting an issue and the possibility to self-heal based on a reliability target over a 30-day period:

Reliability level	Downtime per month	Enough time for issue to be detected by monitoring system?	Possibility that issue can be resolved by?
90%	3.04 days	Yes	Human or machine
99% (*two nines*)	7.30 hours	Yes	Human or machine
99.5%	3.65 hours	Yes	Human or machine
99.9% (*three nines*)	43.83 minutes	Yes	Human or machine
99.99% (*four nines*)	4.38 minutes	Yes	Not enough time for human; possible with a machine
99.999% (*five nines*)	26.30 seconds	Not enough time for issue to be detected by monitoring system	Not enough time for both human and machine

To summarize, a reliability target should be set to a level that is realistic where an issue can be detected and addressed. An automated self-healing process is recommended over a human involvement—for example, redirecting traffic to a new **availability zone (AZ)** in case of an existing AZ failure.

Setting a reliability target too low means that issues could frequently occur, leading to large duration of downtimes, and customers will be impacted regularly. Setting a reliability target too high at 99.999% or even 100% means that the system cannot practically fail, and that makes it difficult to add new features to the service or application.

Setting SLOs is an iterative process

Reliability is the most important feature of a service, and setting SLOs allow monitoring systems to capture how the service is performing. When setting SLOs for the first time, it's possible to set SLOs based on past performance, taking an assumption that users are happy to start with. SLIs for these SLOs are based on existing monitoring systems and are considered as an initial baseline that must be met. Such SLOs are known as **achievable SLOs**, and any misses below the initial base line should result in directing engineering efforts to focus on getting reliability back to the initial baseline.

> **How to get started with setting achievable SLOs**
>
> Metrics to set achievable SLOs can either be taken from the load balancer or backfilled from the logs. Both approaches give an insight into historical performance.

If SLOs need to be set in the absence of historical data or if historical data does not accurately reflect users' expectations, it is recommended to set an achievable target and then refine the target to closely match users' expectations and business needs. Such SLOs are known as **aspirational SLOs**. Monitoring systems will then use these metrics to track these SLOs.

Once either achievable or aspirational SLOs are set, it's possible that new features are introduced to the service, but the probability for a service to be unreliable also increases. This can result in customers being unhappy even after meeting SLOs. This is an indication that monitoring metrics need to be revisited. SLOs need to be iteratively set and periodically re-evaluated. These metrics might have worked when originally set, but might not anymore.

Here are a few possible scenarios that call for SLOs to be re-evaluated:

- New features were not considered in the metric calculation.
- Service usage is now extended from desktop to mobile.
- Service usage is now extended to multiple geographies.

> **How frequently should SLOs be revisited or re-evaluated?**
>
> It's recommended that SLOs be revisited or re-evaluated every 6 to 12 months to ensure that defined SLOs continue to match business changes and users' expectations.

In addition to periodically revisiting SLOs, there are scenarios where a different SLO—more precisely, a tighter SLO—can be used when a spike in traffic is anticipated. For example, during holiday shopping, many businesses expect a significant spike in traffic, and in such scenarios, businesses can come up with a temporary strategy of tightening the SLO from 99.9% to 99.99%. This means system reliability is prioritized over a need or urge to release new features. The SLO targets are set back to their original value (in this example, back to 99.9%) when normal traffic resumes.

This completes the section on SLOs, with a deep insight into the need for reliability, setting reliability targets, and the way SLOs drive business decisions using SLIs. The next subsection introduces why SLOs need SLIs and is also a precursor before exploring SLIs in detail.

SLOs need SLIs

SLOs are specific numerical targets to define the reliability of a system. *SLOs are also used as a prioritization signal* to determine the balance between innovation and reliability. SLOs also help to differentiate happy and unhappy users. But the striking question is: *How do we measure SLOs?*

SLOs are measured using SLIs. These are defined as a quantifiable measure of service reliability and specifically give an indication on how well a service is performing at a given moment of time. Service consumers have certain expectations from a service and SLIs are tied directly to those expectations. Examples of quantifiable SLIs are **latency**, **throughput**, **freshness**, and **correctness**. SLIs are expressed as a percentage of good events across valid events. SLOs are SLI targets aggregated over a period.

We'll get into more details about SLIs in the next section, *Exploring SLIs*. This includes categorizing SLIs by types of user journeys and elaborating on the ways to measure SLIs.

Exploring SLIs

An SLI is a quantitative measure of the level of service provided with respect to some aspect of service reliability. Aspects of a service are directly dependent on potential user journeys, and each user journey can have a different set of SLIs. Once the SLIs are identified per user journey, the next critical step is to determine how to measure the SLI.

This section describes the details around how to identify the right measuring indicators or SLIs by categorizing user journeys, the equation to measure SLIs, and ways to measure SLIs.

Categorizing user journeys

The reliability of a service is based upon the user's perspective. If a service offers multiple features, each feature will involve a set of user interactions or a sequence of tasks. This sequence of tasks that is critical to the user's experience offered by the service is defined as a **user journey**.

Here are some examples of user journeys when using a video streaming service:

- Browsing titles under a specific category—for example, *fiction*
- Viewing an existing title from a user library
- Purchasing an on-demand show or a live stream
- Viewing a live stream

Each user journey can have a different expectation. These expectations can vary, from the speed at which the service responds to a user's request to the speed at which data is processed, to the freshness of the data displayed or to the durability at which data can be stored.

There could be a myriad user journeys across multiple services. For simplicity, user journeys can be classified into two popular categories, as follows:

- **Request/response** user journey
- **Data processing/pipeline**-based user journey

Each category defines specific characteristics. Each specific characteristic can represent an SLI type that defines the reliability of the service. These are specified in the following sections.

Request/response user journey

Availability, latency, and quality are the specific aspects or characteristics of SLIs that need to be evaluated as part of a request/response user journey.

Availability

Availability is defined as the proportion of valid requests served successfully. It's critical for a service to be available to meet user's expectations.

To convert an availability SLI definition into an implementation, a key choice that needs to be made is: *How to categorize requests served as successful?*

To categorize requests served as successful, error codes can be used to reflect users' experiences of the service—for example, searching a video title that doesn't exist should not result in a 500 series error code. However, being unable to execute a search to check if a video title is present or not, should result in a 500 series error code.

Latency

Latency is defined as the proportion of valid requests served faster than a threshold. It's an important indication of reliability when serving user-interactive requests. The system needs to respond within a timely fashion to consider it as interactive.

Latency for a given request is calculated as the time difference between when the timer starts and when the timer stops. To convert a latency SLI definition into an implementation, a key choice that need to be made is: *How to determine a threshold to classify responses as fast enough?*

To determine a threshold that classifies responses as fast enough, it's important to first identify the different categories of user interactions and set thresholds accordingly per category. There are three ways to bucketize user interactions, outlined as follows:

- **Interactive**—Refers to interactions where a user waits for the system to respond after clicking an element. Can also be referred to as **reads**, and a typical threshold is 1,000 ms.

- **Write**—Refers to user interactions that make a change to the underlying service. A typical threshold is 1,500 ms.

- **Background**—Refers to user interactions that are asynchronous in nature. A typical threshold is 5,000 ms.

Quality

Quality is defined as the proportion of valid requests served without degrading the service. It's an important indication on how a service can fail gracefully when its dependencies are unavailable.

To convert a quality SLI into an implementation, a key choice that needs to be made is: *How to categorize if responses are served with degraded quality?* To categorize responses served with degraded quality, consider a distributed system with multiple backend servers. If the incoming request is served by all backend services, then the request is processed without service degradations. However, if the incoming request is processed by all backend servers except one, then it indicates responses with degraded quality.

If a request is processed with service degradation, the response should be marked as degraded, or a counter should be used to increment the count of degraded responses. As a result, a quality SLI can be expressed as a ratio of **bad events** to **total events** instead of a ratio of **good events** to **total events**.

How to categorize a request as valid

To categorize a request as valid, different methodologies can be used. One such method is to use **HyperText Transfer Protocol (HTTP)** response codes. For example, 400 errors are client-side errors and should be discarded while measuring the reliability of the service. 500 errors are server-side errors and should be considered as failures from a service-reliability perspective.

Data processing/pipeline-based user journey

Freshness, correctness, coverage, and throughput are the specific aspects or characteristics of SLIs that need to be evaluated as part of a data processing/pipeline-based user journey. This is also applicable for batch-based jobs.

Freshness

Freshness is defined as the proportion of valid data updated more recently than a threshold. Freshness is an important indicator of reliability while processing a batch of data, as it is possible that the output might become less relevant over a period of time. This is primarily because new input data is generated, and if the data is not processed regularly or rebuilt to continuously process in small increments, then the system output will not effectively reflect the new input.

To convert a freshness SLI into an implementation, a key choice that needs be made is: *When to start and stop the timer to measure the freshness of data?* To categorize that the data processed is valid for SLI calculation, the correct source of input data or the right data processing pipeline job must be considered. For example, to calculate the freshness of weather-streaming content, data from a sports-streaming pipeline cannot be considered. This level of decision making can be achieved by implementing code and a rule-processing system to map the appropriate input source.

To determine when to start and stop times to measure the freshness of data, it is important to include timestamps while generating and processing data. In the case of a batch processing system, data is considered fresh if the next set of data is not processed and generated. In other words, freshness is the time elapsed since the last time the batch processing system completed.

In the case of an incremental streaming system, freshness refers to the age of the most recent record that has been fully processed. Serving stale data is a common way to degrade the response quality. Measuring stale data as degraded response quality is a useful strategy. If no user accesses the stale data, no expectations around the freshness of the data can have been missed. For this to be feasible, one option is to include a timestamp along with generating data. This allows the serving infrastructure to check the timestamp and accurately determine the freshness of the data.

Correctness

Correctness is defined as the proportion of valid data producing a correct output. It's an important indication of reliability whereby processing a batch of data results in the correct output. To convert a correctness SLI into an implementation of it, a key choice that needs to be made is: *How to determine if the output records are correct?*

To determine if the output records produced are correct, a common strategy is to use golden input data, also known as a set of input data that consistently produces the same output. This way, the produced output can be compared to the expected output from the golden input data.

Proactive testing practices—both manual and automated—are strongly recommended to determine correctness.

Coverage

Coverage is defined as the proportion of valid data processed successfully. It's an important indication of reliability, whereby the user expects that data will be processed and outputs will subsequently also be available.

To convert a coverage SLI into an implementation, the choice that needs to be made is: *How to determine that a specific piece of data was processed successfully?* The logic to determine if a specific piece of data was processed successfully should be built into the service, and the service should also track the counts of success and failure.

The challenge comes when a certain set of records that were supposed to be processed are skipped. The proportion of the records that are not skipped can be known by identifying the total number of records that should be processed.

Throughput

Throughput is defined as the proportion of time where the data processing rate is faster than a threshold. It's an important indicator of reliability of a data processing system, whereby it accurately represents user happiness and operates continuously on streams or small batches of data.

To convert a throughput SLI into an implementation, a key choice that needs to be made is: *What is the unit of measurement for data processing?* The most common unit of measurement for data processing is **bytes per second (B/s)**.

It is not necessary that all sets of inputs have the same throughput rate. Some inputs need to be processed faster and hence require higher throughput, while some inputs are typically queued and can be processed later.

> **SLIs recommended for a data storage-based user journey**
>
> Systems processing data can also be further classified into systems responsible for only storing data. So, a **data storage** user-based journey is another possible classification of a user journey where availability, durability, and end-to-end latency are additional recommended SLIs. Availability refers to data that could be accessed on demand from a storage system. **Durability** refers to the proportions of records written that could be successfully read from a storage system as and when required at that moment. **End-to-end latency** refers to the time taken to process a data request, from ingestion to completion.

The following table summarizes specific characteristics to represent an SLI type, grouped by the type of user journey:

User journey type	Characteristics to represent SLI
Request/response user journey	Availability, latency, quality
Data processing/pipeline-based user journey	Freshness, correctness, coverage, throughput
Data storage user journey	Availability, durability, and end-to-end latency

Given that there is a wide choice of SLIs to select from, Google recommends the following specific SLIs based on the type of systems:

- **User-facing serving systems**: Availability (*Is it possible to respond to a request?*), latency (*How long will it take to respond?*), and throughput (*How many requests can be handled?*)

- **Storage systems**: Latency (*How long does it take to read or write data?*), availability (*Can the data be accessed on demand?*), and durability (*Is the data still available when there is a need?*)

- **Big data systems**: Throughput (*How much data can be processed?*) and end-to-end latency (*What is the time duration for data to progress from ingestion to completion?*)

Given that we have looked at various factors that impact on determining SLIs specific to a user journey, the upcoming subsection will focus on the methodology and sources to measure SLIs.

SLI equation

An SLI equation is defined as the proportion of valid events that were good, as illustrated here:

$$SLI = \frac{good\ events}{valid\ events} * 100$$

This equation has the following properties:

- SLIs are expressed as a percentage and fall between 0% and 100%. 100% refers to everything working, and 0% refers to everything being broken.
- SLIs consistently translate to percentage reliability, SLOs, and error budgets, and are also key inputs to **alerting logic**.
- SLIs allow us to build common tooling to reflect the reliability of a service or system.
- Valid events are determined as follows: for requests related to **HTTP Secure** (**HTTPS**), valid events are determined based on request parameters or response handlers. Request parameters can include hostname or request path. For requests related to data processing systems, valid events refer to the selection of specific inputs that scope to a subset of data.

This completes our summary of SLI equation and its associated properties. The next subsection details various popular sources to measure SLIs.

Sources to measure SLIs

Identifying potential user journeys for a service is the first important step to identify SLIs. Once SLIs to measure are identified, the next key step is to measure the SLIs so that corresponding alerts can be put in place. The key question in this process is: *How to measure and where to measure?*

There are five popular sources or ways to measure SLIs, outlined as follows:

- Server-side logs
- Application server
- Frontend infrastructure
- Synthetic clients
- Telemetry

Server-side logs

Here are some details on how information from server-side logs can be used to measure SLIs:

- Logs capture multiple request-response interactions over a long-running session. Stored logs give an option to get insights into historical performance of the service.

- If starting out with setting SLOs (*SLOs need SLIs*), for a service, log data can be used to analyze historical events, reconstruct user interactions, and retroactively backfill the SLI information.

- Complex logic can be added to the code itself where good events are clearly identified, and the information is captured in logs. (*This requires significant engineering efforts.*)

Here are details of the limitations of using server-side logs to measure SLIs:

- If an SLI needs to be used to trigger an emergency response, the time between the event occurrence and the event actually being measured should be minimal. Given that logs need to be ingested and processed, capturing SLIs from logs will add significant latency.

- Log-based SLIs cannot capture the requests that did not make it to the application server.

Application server

Here are details on how information from the application server can be used to measure SLIs:

- Metrics captured at the application server are known as application-level metrics. These metrics are helpful in diagnosing issues with respect to the application.

- Application metrics can capture the performance of individual requests without measurement latency. In addition, these events could be aggregated over time.

Here are details of the limitations of using the application server to measure SLIs:

- Application metrics cannot capture complex multi-request user journeys.

- Application-based SLIs cannot capture requests that do not make it to the application server.

> **What is a complex multi-request user journey?**
>
> A complex multi-request user journey will include a sequence of requests, which is a core part of a user consuming a service such as searching a product, adding a product to a shopping cart, and completing a purchase. Application metrics cannot capture metrics for the user journey but can capture metrics related to individual steps.

Frontend infrastructure

Here are details of how information from frontend infrastructure can be used to measure SLIs:

- Frontend infrastructure refers to load balancers. This could be a vendor-based load balancer (such as *F5*) or a cloud-provider based load balancer (such as *Google Cloud Load Balancer*).

- Most of the distributed applications use a load balancer, and this is the first point of interaction for a user's request before it is sent to the actual application. This makes the load balancer the closest point to the user and fewer requests go unmeasured.

- Cloud providers typically capture multiple metrics for the incoming requests to the load balancer out of the box. This information might be readily available, including historical data too. If the capture of data was not configured for some reason, it can be easily configured. Either way, information will be available without investing engineering efforts (*as compared to capturing metrics from application logs*).

- Load balancers capture metrics related to requests that do not make it to the application server.

Here are details of the limitations of using a frontend infrastructure to measure SLIs:

- Load balancers can either be **stateful** or **stateless**. If stateless, then the load balancers cannot track user sessions and hence cannot be used to capture metrics tied to user interactions.

- Given that load balancers typically act as a **traffic cop**, routing user requests to application servers that are capable of handling the requests, load balancers do not inherently have control over the response data returned by the application. Instead, load balancers are dependent on the application to set the metadata accurately on the response envelope.

- The dependency on the application to set the right metadata on the response envelope is a conflict of interest because it is the same application that is generating the metrics.

Synthetic clients

Here are details of how information from synthetic clients can be used to measure SLIs:

- Synthetic clients provide **synthetic monitoring**, a monitoring technique that monitors the application by emulating or simulating user interactions based on a recorded set of transactions.

- Synthetic clients can emulate user interactions that constitute a user journey from a point outside the infrastructure, and hence can verify the responses.

Here are details of the limitations of using synthetic clients to measure SLIs:

- Synthetic clients simulate a user's behavior, and hence it's an approximation.

- Synthetic clients need complex integration tests that could cover multiple edge cases, thus resulting in a significant engineering effort.

- Synthetic clients need maintenance to add new user simulations if new user behavior patterns emerge that were not previously accounted for.

Telemetry

Telemetry refers to remote monitoring from multiple data sources and is not restricted to capture metrics related to application health, but can be extended to capture security analytics such as suspicious user activity, unusual database activity, and so on. Here are details of how information from telemetry can be used to measure SLIs:

- Instrumenting clients to implement telemetry metrics helps to measure the reliability of third-party integration systems such as a **content delivery network** (**CDN**).

- **OpenTelemetry** is the most popular instrumentation mechanism to capture traces and metrics. It replaces **OpenTracing** and **OpenCensus**, which were individually focused on capturing tracing and metrics respectively.

What is OpenTelemetry?

OpenTelemetry is a unified standard for service instrumentation. It provides a set of **application programming interfaces** (**APIs**)/libraries that are vendor-agnostic and standardizes how to collect and send data to compatible backends. OpenTelemetry is an open source project that is part of the **Cloud Native Computing Foundation** (**CNCF**).

Here are details of the limitations of using telemetry to measure SLIs:

- Ingesting metrics from different sources increases latency and will pose the same issues encountered when capturing metrics from processing logs, thus this is not a good fit for triggering emergency responses.

- If telemetry is implemented in-house, it requires a significant engineering effort. However, there are vendors that provide the same capability, but there is a risk of vendor lock-in.

This completes an elaboration of five different sources to measure SLIs. Given that each source has its own limitations, there is no best source to measure SLIs. In most cases, a combination of sources is always preferred. For example, if an organization is getting started with their SRE practice, usage of server-side logs to backfill SLIs and frontend infrastructure to readily use the metrics from the load balancer might be a good way to start. It can later be extended to capturing metrics from the application server, but given it doesn't support complex multi-request user journeys, an organization can later shift to the use of telemetry or synthetic clients based on their need. The next subsection summarizes a few SLI best practices as recommended by Google.

SLI best practices (Google-recommended)

It's a tedious task for an organization that would like to start on their SRE journey, and a key aspect of this journey is to identify, define, and measure SLIs. Here is a list of Google-recommended best practices:

- **Prioritize user journeys**: Select user journeys that reflect features offered by the service and the user's affinity to those features.

- **Prioritize user journeys from the selected list**: The user journey to purchase or watch a streaming event is more important than rating a video.

- **Limit number of SLIs**: Keep it to three to five SLIs per user journey. More SLIs will make it complex to manage for the operators.

- **Collect data via frontend infrastructure**: Collecting at load balancer level is closer to a user's experience and requires less engineering effort.

- **Aggregate similar SLIs**: Collect data over a period. Convert metric information captured into rate, average, or percentile.

- **Keep it simple**: Complex metrics require significant engineering effort but might also increase response time. If response time increases, then the metric will not be suitable for emergency situations.

This completes a comprehensive deep dive on SLIs, with a focus on categorizing user journeys, identifying specific aspects that impact a user journey, various sources to measure, and recommended best practices to define SLIs. To summarize, there are four critical steps for choosing SLI, listed as follows:

1. Choose an SLI specification based on a suitable user journey.
2. Refine a specification into a detailed SLI implementation.
3. Walk through the user journey and identify implementation gaps.
4. Set aspirational SLO targets based on business needs.

The upcoming section focusses on error budgets, which are used to achieve reliability by maintaining a balance with release velocity.

Understanding error budgets

Once SLOs are set based on SLIs specific to user journeys that define system availability and reliability by quantifying users' expectations, it is important to understand how unreliable the service is allowed to be. This acceptable level of unreliability or unavailability is called an **error budget**.

The unavailability or unreliability of a service can be caused due to several reasons, such as planned maintenance, hardware failure, network failures, bad fixes, and new issues introduced while introducing new features.

Error budgets put a quantifiable target on the amount of unreliability that could be tracked. They create a common incentive between development and operations teams. This target is used to balance the urge to push new features (*thereby adding innovation to the service*) against ensuring service reliability.

An error budget is basically the inverse of availability, and it tells us how unreliable your service is allowed to be. If your SLO says that 99.9% of requests should be successful in a given quarter, your error budget allows 0.1% of requests to fail. This unavailability can be generated because of bad pushes by the product teams, planned maintenance, hardware failures, and other issues:

$Error\ Budget = 100\% - SLO$

Here's an example. If SLO says that 99.9% of requests should be successful in a given quarter, then 0.1% is the error budget.

Let's calculate the error budget, as follows:

If SLO = 99.9%, then error budget = 0.1% = 0.001

Allowed downtime per month = 0.001 * 30 days/month * 24 hours/day * 60 minutes/hour =
43.2 minutes/month

This introduces the concept of an error budget. The next subsection introduces the concept of an error budget policy and details the need for executive buy-in with respect to complying with the error budget policy.

Error budget policy and the need for executive buy-in

If reliability is the most important feature of a system, an **error budget policy** represents how a business balances reliability against other features. Such a policy helps a business to take appropriate actions when the reliability of the service is at stake. The key to defining an error budget policy is to actually decide the SLO for the service. If the service is missing SLO targets, which means the error budget policy is violated, then there should be consequences. These consequences should be enforced by generating executive buy-in. Operations teams should have an influence on the impact of the development team's practices by halting the release of new features if the service is getting very close to exhausting the error budget or has exceeded the error budget.

Error budgets can be thought of as funds that are meant to be spent across a given time period. These funds can be spent on releasing new features, rolling out software updates, or managing incidents. An error budget is basically the inverse of availability, and it tells us how unreliable your service is allowed to be. If your SLO says that 99.9% of requests should be successful in a given quarter, your error budget allows 0.1% of requests to fail. This unavailability can be generated because of bad pushes by the product teams, planned maintenance, hardware failures, and so on. The next subsection lists out the characteristics of an effective error budget policy.

Characteristics of an effective error budget policy

An error budget policy should have the following characteristics:

- An overview of the service
- A list of intended goals
- A list of non-goals; also referred to as a potential requirement that has been specifically excluded
- A list of *DOs* and *DON'Ts* based on whether a service performs above its SLO or misses its SLO
- A list of consequences when a service misses its SLO

- A detailed outage policy that defines the criteria to call out an incident and a need for a follow-up to ensure the incident doesn't happen again

- A clearly laid-out escalation policy that identifies the decision maker in the event of a disagreement

The preceding list of characteristics clearly calls out the fact that it is extremely necessary to have well-documented SLOs to define an effective error budget policy. This will be discussed in the next subsection.

Error budgets need well-documented SLOs

The key to defining an error budget is to actually to decide the SLOs for the service. SLOs clearly differentiate between reliable services and unreliable services, thus extending it to identify happy versus unhappy users. SLOs should be clearly defined without any ambiguity and should be agreed by product owners, developers, SREs, and executives.

In addition to implementing an SLO and configuring a monitoring system to alert on the SLO, the following characteristics are recommended for a well-documented SLO in terms of metadata:

- The need for an SLO and thought process behind the specific SLO target

- A list of owners for SLOs

- The impact in case of SLO miss

- SLIs tied with the SLOs

- Any specific events that are included or excluded from the calculation

- Version control of the SLO documentation (*this gives an insight into reasons for changing SLOs as they get refined over a period*)

The next subsection discusses multiple options to set error budgets.

Setting error budgets

Error budgets can be thought as funds that are meant to be spent across a given time period. These funds can be spent on releasing new features or rolling out software updates or managing incidents. But this raises several questions, such as the following:

- What is the right time to spend error budgets? At the start of the month or the end of the month?

- What happens if the error budget gets exhausted and there is an emergency?

- What happens if the error budget is not exhausted? Can it be carried over?

Different strategies can be used to determine the right time to spend error budgets within a time period. Let's assume the time period is 28 days. There could be three potential options, listed as follows:

- **Option #1** is to spend a portion of the error budget at the start (of the 28 days) to push new features or updates and use the remaining error budget for potential reliability maintenance in the event of an incident.

- **Option #2** is to spend an error budget to push new features or updates after elapsing half of the time period (say, 14 days), since it gives an idea on how much error budget was used in the first half for maintaining system reliability.

- **Option #3** is to spend any remaining error budget toward the latter part of the time period in pushing new features or updates to ensure focus is on system reliability till then.

Any of the preceding options or a combination of the three can be used to define a dynamic release process, and it all depends on what developers and operations team agree upon based on current business needs and past performance. The dynamic release process can be implemented by setting alerts based on error budget exhaustion rates.

If the error budget of a service is exhausted but the development team needs to push a new feature as an exception scenario, SRE provisions this exception using **silver bullets**.

Envision silver bullets as tokens that could be given to the operations team to facilitate an exception to release new features when having exceeded the error budget. These tokens reside with a senior stakeholder and the development team needs to pitch the need to use silver bullets to the stakeholder. A fixed number of such tokens are given to the stakeholder and these are not carried over to the next time period. In addition to the use of silver bullets, SRE also recommends the use of rainy-day funds whereby a certain amount of the error budget is additionally provided to handle unexpected events.

Error budgets cannot be carried over to the next time period. So, in all practicality, the goal is to spend the error budget by the end of the time period. Constantly exhausting error budgets and repeated use of silver bullets should call for a review, where engineering efforts should be invested in making the service more reliable by improving the service code and by adding integration tests.

The use of dynamic release cadence, error budget exhaustion rates, silver bullets, and rainy-day funds are advanced techniques prescribed by SRE to manage error budgets. This completes the subsection on defining characteristics for an effective error budget policy, listing out characteristics for well-documented SLOs and discussing options to set error budgets. The next subsection details factors that are critical in ensuring that a service stays reliable and does not exhaust the error budget.

Making a service reliable

When a service exhausts its error budget or repeatedly comes close to exhausting the same, engineering teams should focus on making a service reliable. This raises the next obvious question: *How can the engineering teams make a service more reliable to meet users' expectations?*

To get deeper insights into this, it's critical to consider the following key factors essential to determine the potential impact on the service:

- **Time to detect** (**TTD**)—Defined as the difference in time from when the issue first occurred to the time that the issue was first observed or reported. Example: If an issue occurred at 10 a.m. but was reported or observed at 10:30 a.m., then the TTD in this scenario is 30 minutes.

- **Time to resolve** (**TTR**)—Defined as the difference in time from when the issue was first observed or reported to the time that the issues was resolved. Example: If an issue was first observed or reported at 10:30 a.m. but was resolved at 10:45 a.m., then the TTR in this scenario is 15 minutes.

- **Time to fail** (**TTF**)—Defined as how frequently the service is expected to fail. TTF is also known as **Time Between Failures**, or **TBF**.

- **Impact %**—Percentage of impact in terms of impacted users or impacted functional areas.

- The expected impact is proportional to the following expression:

$$Expected\ Impact\ \approx\ \frac{(TTD + TTR) * impact\ \%}{TTF} * 100$$

Reliability can be improved by implementing the following options:

- Reducing detection time: Reduce TTD

- Reducing repair time: Reduce TTR

- Reduce impact %: Reduce the impacted users/functional areas

- Reduce frequency: Increasing TTF

- Operational improvement

Let's discuss each option in detail, next.

Reduce detection time (TTD)

TTD can be reduced by the following approaches:

- Add automated alerting that alerts a user rather than the user manually detecting an issue by noticing an abnormality from the metrics dashboard.

- Add monitoring to measure SLO compliance. This helps to know how quickly the error budget is being consumed or whether the service is performing within its target SLOs.

Reduce repair time (TTR)

TTR can be reduced by the following approaches:

- Develop a playbook that makes it easier to parse and collate server debug logs. This will help the onsite engineers to quickly address the problem at hand. If a new pattern was detected, then the playbook should be updated.

- Automate manual tasks such as increasing disk space to an acceptable percentage of current disk space, draining zone or rerouting traffic.

- Collect relevant information for a specific scenario, which will save time for the on-call team and will allow them to get a head start with their investigation.

Reduce impact %

Impact % can be reduced by the following approaches:

- Roll out a new feature to a limited number of users through a percentage-based rollout within a given amount of time. This reduces the impact percentage to that specific user group. The rollout percentage can be gradually increased from 0.1% of users to 1%, then later to 10%, and eventually 100%. In this way, the releases are staggered.

- Engineer a service to run in a degraded mode during a failure, such as switching to allow read-only actions and not writes, thus reducing the impact.

Reduce frequency

Frequency can be reduced by the following approaches:

- Run the service or application in multiple zones or regions. Direct the traffic away from the zone or region that failed to an alternate working zone or region.

- Create a self-healing automated script that reduces the impact and frequency, but also report the issue so that it can be addressed later.

Operational improvements

Here are some options from an operational standpoint to make a service reliable:

- Increase availability by adding redundancy and thereby remove single points of failure.

- Identify a common category between failures that can point to a specific region or a specific set of customers who consume a majority of the error budget.

- Standardize the infrastructure or minimize the differences to achieve similar results when testing a service against running it in production.

- Use design patterns that allow the service to be rolled back in case of an issue.

- Create alerts by tracking the error budget burndown rate.

- Use post-mortems to identify the issues at hand and create actionable items to fix those issues.

This completes a complete deep dive into potential factors that needs to be considered and feasible options that can be implemented to make a service reliable, thus not consuming the error budget. The next subsection summarizes the section on error budgets.

Summarizing error budgets

Error budgets can be summarized by the following key pointers:

- Error budgets help development and SRE teams balance release velocity and stability. Management buys into SLO and gives executive backing.

- Error budgets need SLOs, and SLOs need SLIs to monitor service reliability.

- The difference between the actual and targeted SLO is calculated to determine if it is below the error budget, and if so, the release of new features is allowed.

- If not, engineering efforts should be focused on the reliability of the service.

- An error budget policy is an effective way to implement the concepts of an error budget.

This completes the section on error budgets, with a deep dive into multiple aspects that include how to define an effective error budget policy, how to set error budgets, the impact of having an executive buy-in that helps to make a service reliable, and how to effectively balance the release velocity of new features.

Eliminating toil through automation

Toil was introduced in *Chapter 1, DevOps, SRE, and Google Cloud Services for CI/CD*, and is defined as the work tied to a production service where the characteristic of that work is manual, repetitive, automatable, tactical, lacks enduring value, and linearly grows with the service. Toil is often confused with overhead, but overhead refers to administrative work that includes email, commute, filing expense reports, and attending meetings. Toil can be both good and bad—it really depends on the amount of toil.

Here are some of the positive sides of performing toil, but in very short and limited amounts:

- Produces a sense of gratification or a sense of accomplishment
- Can act as a low-stress or low-risk activity
- Can be used to train new recruits, especially providing them a chance to learn by being hands-on with the system to learn the inner workings

However, excessive toil can lead to the following problems or issues:

- **Career stagnation**: Solving production issues is gratifying, but solving the same issue manually in a repetitive manner doesn't help from a career standpoint. This takes away learning time and intent away from the SRE engineer, leading to stagnation.

- **Burnout and boredom**: Excessive toil leads to burnout and boredom. An SRE engineer will be bored doing the same task every day. Sometimes, manual tasks might also be tedious and laborious, leading to burnout.

- **Low feature velocity**: If the SRE team is engaged in lot of toil, then they will have less time to work on releasing new features, thus leading to lower feature velocity and reduced release cadence.

- **Wrong precedence**: If the SRE team engages in more toil than required, it's possible that development teams will make SRE teams do further toil, especially on items that need to be addressed truly by the development team instead to remove the root cause. This will create confusion on the role of the SRE engineer.

All the aforementioned problems or issues can potentially lead to attrition, as SRE engineers might not be happy with their everyday work and might look elsewhere for better work and challenges. SRE recommends that toil should be bounded and that an SRE engineer should not work more than 50% of their time on toil. Anything more than 50% blurs the line between an SRE engineer and a system administrator. SRE engineers are recommended to spend the remaining 50% on supporting engineering teams in achieving reliability goals for the service.

Eliminating toil allows SRE engineers to add service features to improve the reliability and performance of the service. In addition, focus can continue to remain on removing toil as identified, thus clearing out a backlog of any manual repetitive work. SRE encourages the use of engineering concepts to remove manual work. This also allows SRE engineers to scale up and manage services better than a development or an operations team.

SRE recommends removing toil through automation. Automation provides consistency and eliminates the occurrence of oversights and mistakes. Automation helps to perform a task much faster than humans and can also be scheduled. Automation also ensures to prevent a problem before reoccurring. Automation is usually done through code, and this also provides a chance for SRE engineers to use engineering concepts to implement the required logic.

This concludes the section on toil: its characteristics, the good and bad aspects, and the advantages of using automation to eliminate toil. The next section illustrates how SLAs, SLOs, and error budgets are impacted based on SLI performance.

Illustrating the impact of SLAs, SLOs, and error budgets relative to SLI

In this section, we will go through two hands-on scenarios to illustrate how SLO targets are met or missed based on SLI performance over time. SLOs performance will have direct impact on SLAs and error budgets. Changes in the error budget will specifically dictate the priority between the release of new features versus service reliability. For ease of explanation, a 7-day period is taken as the measure of time (ideally, a 28-day period is preferred).

Scenario 1 – New service features introduced; features are reliable; SLO is met

Here are the expectations for this scenario:

- **Expected SLA**—95%
- **Expected SLO**—98%
- **Measured SLI**—Service availability or uptime
- **Measure duration**—7 days

Given that the anticipated SLO for service is 98%, here is how the allowed downtime or error budget is calculated (you can use this downtime calculator for reference: `https://availability.sre.xyz`):

- **Error budget** = *100% - SLO = 100% - 98% = 2%*
- **Allowed downtime for 7 days with 98% availability** = *3.36 hours = 3.36 * 60 = 201.6 minutes*

So, if total downtime across 7 days is less than 201.6 minutes, then the service is within SLO compliance of 98%, else the service is out of SLO compliance.

Now, let's illustrate how the SLO is impacted based on SLI measurements. Assume that new features are introduced for the service (across the 7-day period) and the features are stable, with minimal issues.

The following table represents the SLI measurements of availability, respective downtime based on SLI performance, and the reduction in error budget on a per-day basis:

Day #	SLI performance	Observed downtime (minutes)	Remaining error budget (minutes)
1	99.0 %	14.40	201.60 - 14.40 = 187.2
2	98.9 %	15.84	187.20 - 15.84 = 171.36
3	99.2 %	11.52	171.36 - 11.52 = 159.84
4	99.2 %	11.52	159.84 - 11.52 = 148.32
5	99.0 %	14.40	148.32 - 14.40 = 131.04
6	98.8 %	17.28	131.04 - 17.28 = 113.76
7	98.0 %	28.80	113.76 - 28.80 = 84.95

The following screenshot represents the SLI performance for service uptime (top) and the error budget burndown rate (bottom) based on the values from the preceding table:

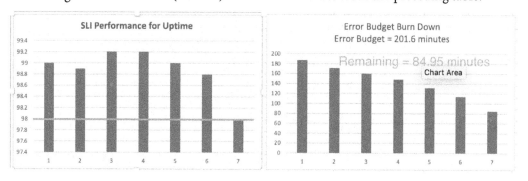

Figure 2.2 – Illustration of SLI performance and error budget burndown rate

Here are some critical observations:

- Remaining error budget = 84.95 minutes, which is less than 201.6 minutes. Hence, the SLO is in compliance and the SLA is met.

- The SLO performance can be calculated based on the remaining error budget. Since the remaining error budget is 84.95 minutes, then SLO performance is 99.25%.

- The introduction of new features did not unevenly or suddenly decrease the error budget.

This completes a detailed illustration of a scenario where the SLO is met based on the SLI performance over a 7-day period. The next scenario illustrates the opposite, where the SLO is missed based on SLI performance.

Scenario 2 – New features introduced; features are not reliable; SLO is not met

Here are the expectations for this scenario:

- **Expected SLA**—95%

- **Expected SLO**—98%

- **Measured SLI**—Service availability or uptime

- **Measure duration**—7 days

As calculated in *Scenario 1*, the allowed downtime for a 98% SLO is 201.6 minutes. So, the SLO is out of compliance if downtime is greater than 201.6 minutes over a 7-day period.

Now, let's illustrate how the SLO is impacted based on SLI measurements. Assume that new features are introduced for the service (across the 7-day period) but the introduced features are not stable, causing major issues resulting in longer downtimes.

The following table represents the SLI measurements of availability, respective downtime based on SLI performance, and the reduction in error budget on a per-day basis:

Day #	SLI performance	Observed downtime (minutes)	Remaining error budget (minutes)
1	98.0 %	28.80	201.60 - 28.80 = 172.80
2	97.4 %	37.44	172.80 - 37.44 = 135.36
3	94.5 %	79.20	135.36 - 79.20 = 56.16
4	95.5 %	64.80	0
5	98.0 %	28.80	0
6	99.5 %	7.20	0
7	99.5 %	7.20	0

The following screenshot represents the SLI performance for service uptime (left-hand side) and the error budget burndown rate (right-hand side) based on the values from the preceding table:

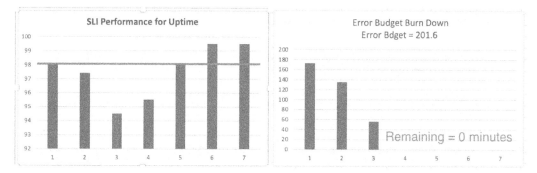

Figure 2.3 – Illustration of SLI performance and error budget burndown rate

Here are some critical observations:

- Remaining error budget = 0 minutes on day 4. So, the SLO is not in compliance.

- Total downtime across 7-day period is 253.44 minutes.

- The corresponding SLO performance is approximately at 97.48, which is below 98%.

- The SLA is violated since the SLO is not in compliance. This will result in consequences.

- The SRE team should not add new features after day 3 and instead focus on system reliability.

This brings an end to a detailed rundown of *Scenario 2*. This completes our illustration of how SLAs, SLOs, and error budgets are impacted based on SLI performance. This also means we have reached the end of this chapter.

Summary

In this chapter, we discussed in detail the key SRE technical practices: SLAs, SLOs, SLIs, error budgets, and eliminating toil. This included several critical concepts such as factors that can be used for a well-defined SLA, providing guidelines to set SLOs, categorizing user journeys, detailing sources to measure SLIs along with their limitations, elaborating on error budgets, detailing out factors that can make a service reliable, understanding toil's consequences, and elaborating on how automation is beneficial to eliminate toil. These concepts allow us to achieve SRE's core principle, which is to maintain the balance between innovation and system reliability and thus achieve the eventual goal: *build reliable software faster*.

In the next chapter, we will focus on concepts required to track SRE technical practices: monitoring, alerting, and time series. These concepts will include monitoring as a feedback loop, monitoring sources, monitoring strategies, monitoring types, alerting strategies, desirable characteristics of an alerting system, time-series structures, time-series cardinality, and metric types of time-series data.

Points to remember

Here are some important points to remember:

- 100% is an unrealistic reliability target.
- Log-based SLIs and ingesting telemetry adds latency.
- App metrics are not good for complex use journeys.
- SLOs must be set based on conversations with engineering and product teams.
- If there is no error budget left, the focus should be on reliability.
- TTD is the time taken to identify that an issue exists or is reported.
- TTR is the time taken to resolve an issue.
- To improve the reliability of a service, reduce TTD, reduce TTR, reduce impact %, and increase TTF/TBF.
- SLIs should have a predictable relationship with user happiness and should be aggregated over time.
- User expectations are strongly tied to past performance.
- Setting values for SLIs and SLOs should be an iterative process.
- Advanced techniques to manage error budgets are dynamic release cadence, setting up error budget exhaustion rates, rainy-day funds, and the use of silver bullets.
- Identify repetitive tasks that contribute to toil and automate them.

Further reading

For more information on **Google Cloud Platform's (GCP's)** approach toward DevOps, read the following articles:

- **SRE**: `https://landing.google.com/sre/`
- **SRE fundamentals**: `https://cloud.google.com/blog/products/gcp/sre-fundamentals-slis-slas-and-slos`

- **SRE YouTube playlist**: `https://www.youtube.com/watch?v=uTEL8Ff1Zv`
  `k&list=PLIivdWyY5sqJrKl7D2u-gmis8h9K66qoj`

Practice test

Answer the following questions:

1. Which from the following indicates work that is not tied to a production service?

   a) Toil

   b) Manual

   c) Overhead

   d) Automation

2. Which of the following represents *an explicit or implicit contract with your users that includes consequences of meeting or missing the SLOs?*

   a) SLI

   b) SLO

   c) SLA

   d) Error budget

3. Which of the following combinations represent metrics that are typically tracked as part of a request/response journey?

   a) Availability, latency, and durability

   b) Latency, coverage, throughput, and availability

   c) Coverage, correctness, and quality

   d) Availability, latency, and quality

4. Select an option that represents SRE recommendation in terms of the time that an SRE engineer is allowed to spend on toil:

   a) 25%-55%

   b) 45%-60%

   c) 50%-75%

   d) 30%-50%

5. Which of the following is the least realistic (preferred) option to target reliability for a service as an SLO?

    a) 99.9%

    b) 99.99%

    c) 99.999%

    d) 100%

6. In terms of best practice, which option is correct with respect to the number of SLIs recommended per user journey?

    a) 2 to 3

    b) No specific limit

    c) 3 to 5

    d) 5 to 7

7. An e-commerce web application is processing customer purchases through requests and storing the sales transactions in a database. The goal is to ensure that the forecasted sales numbers are based on the latest sales numbers. Which of the following should be selected as SLIs for the e-commerce application (*select two*)?

    a) Database—Availability.

    b) Database—Durability.

    c) Database—Freshness.

    d) Web application—Availability.

    e) Web application—Durability.

    f) Both the database and web application should be available. Production apps should have full availability.

8. Which out of the following represents a precise numerical target for system availability?

    a) SLA

    b) SLO

    c) SLI

    d) Error budget

9. Which of the following represents a direct measurement of service behavior?

    a) SLA

    b) SLO

    c) SLI

    d) Error budget

10. Which of the following is the best suitable source to backfill an SLI?

    a) Application server

    b) Frontend infrastructure

    c) Synthetic clients

    d) None of the above

Answers

1. (c) Overhead.

2. (c) SLA.

3. (d) Availability, latency, and quality.

4. (d) 30%-50%. The recommended amount of toil should not exceed 50%.

5. (d) 100%.

6. (c) 3 to 5.

7. (a) and (d): Both the database and web application should be available as these are production applications.

8. (b) SLO.

9. (c) SLI.

10. (d) None of the above. The exact answer is server-side logs.

3

Understanding Monitoring and Alerting to Target Reliability

Reliability is the most critical feature of a service or a system. **Site Reliability Engineering (SRE)** prescribes specific technical tools or practices that help measure characteristics to define and track reliability, such as **SLAs**, **SLOs**, **SLIs**, and **Error Budgets**. *Chapter 2, SRE Technical Practices – Deep Dive*, took a deep dive into these SRE technical practices across multiple topics, including a blueprint for a well-defined SLA, the need for SLOs to achieve SLAs, the guidelines for setting SLOs, the need for SLIs to achieve SLOs, the different types of SLIs based on user journey categorization, different sources to measure SLIs, the importance of error budgets, and how to set error budgets to make a service reliable.

SLAs are external promises made to the customer, while SLOs are internal promises that need to be met so that SLAs are not violated. This raises a raft of important questions:

- How to observe SLAs for a service so that user or customer expectations are met

- How to observe SLOs for SLAs so that the service is reliable, and SLAs are met

- How to observe SLIs for SLOs so that the service is reliable, and SLAs are met

The preceding questions are critical because it not only has an impact on a user's expectations regarding the service, but also leads to an imbalance between development velocity to deliver new features versus system reliability. This ultimately impacts the promised SLAs, leading to financial or loyalty repercussions. So, in simple terms, the main goal is to identify how to track SRE technical practices to target system reliability.

To track SRE technical practices, three fundamental concepts are required: **Monitoring**, **Alerting**, and **Time Series**. Monitoring is the process of monitoring key indicators that represent system reliability. Alerting is the process of alerting or reporting when the key indicators monitored fall below an acceptable threshold or condition. Monitoring and alerting are configured as a function of time. This means that the data needs to be collected at successive, equally spaced points in time representing a sequence of discrete-time data. This sequence of discrete-time data is also known as a time series.

In this chapter, we're going to explore the following topics and their role in relation to target system reliability:

- **Monitoring**: Feedback loop, monitoring types, and golden signals?

- **Alerting**: Key attributes and approaches for an alerting strategy?

- **Time series**: Structure, cardinality, and metric types?

Understanding monitoring

Monitoring is defined by Google SRE as the action of collecting, processing, aggregating, and displaying real-time quantitative data about a system, such as query counts and types, error counts and types, processing times, and server lifetimes.

In simple terms, the essence of monitoring is to verify whether a service or an application is behaving as expected. Customers expect a service to be reliable and delivering the service to the customer is just the first step. But ensuring that the service is reliable should be the desired goal. To achieve this goal, it is important to explore key data, otherwise also known as metrics. Examples of some metrics can be tied to uptime, resource usage, network utilization, and application performance.

Monitoring is the means of exploring metric data and providing a holistic view of a system's health, which is a reflection of its reliability. Apart from metric data, monitoring can include data from text-based logging, event-based logging, and distributed tracing. The next topic details how monitoring acts as a continuous feedback loop that is critical to continuously improving system reliability by providing constant feedback.

Monitoring as a feedback loop

The ultimate goal is to build a reliable system. For a system to be reliable, the system needs to be continuously observed, in terms of understanding the system's internal states based on its external outputs. This process is known as **observability**.

Observability helps to identify performance bottlenecks or investigate why a request failed. But for a system to be observable, it is important to collect and track several sources of outputs related to the health of the application. These outputs give insights into the application's health and identify any outstanding problems. This is referred to as monitoring. So, monitoring provides inputs that help a system to be observable. In simple terms, *monitoring indicates when something is wrong, while observability helps to show why something went wrong.*

Once an application is deployed, there are four primary areas across which the application needs to be inspected or otherwise monitored:

- Verify an application's performance against the application objectives and identify any deviations in performance by raising relevant questions.

- Analyze data collected by the application over a period of time.

- Alert key personnel when key issues are identified through insights or data analysis.

- Debug the captured information to understand the root cause of an identified problem.

These areas or categories provide a **continuous feedback loop** as part of monitoring the system. This feedback helps to continuously improve the system by identifying issues, analyzing the root cause, and resolving the same. Each of the four categories mentioned is elaborated on in this chapter to provide further insights.

One of the key aspects of monitoring is to raise relevant questions pertaining to the health of the system. This is covered as the next topic.

Raising relevant questions

It is important to raise relevant questions to monitor a system's health post-deployment. These questions provide feedback on how the system is performing. Here are some questions to define where monitoring the service/system can effectively provide a feedback loop:

- Is the size of the database growing faster than anticipated?
- Has the system slowed down after taking the latest software update of a specific system component?
- Can the use of new techniques aid system performance (such as the use of **Memcached** to improve caching performance)?
- What changes are required to ensure that the service/system can accept traffic from a new geographic location?
- Are traffic patterns pointing to a potential hack of the service/system?

The key to answering the preceding questions is to analyze the data at hand. The next topic introduces the possibilities as a result of data analysis.

Long-term trend analysis

Data analysis always leads to a set of trends or patterns. Monitoring these trends can lead to one or more of the following possibilities:

- Point to an existing issue.
- Uncover a potential issue.
- Improve system performance to handle a sudden period of increased traffic.
- Influence experimentation of new system features to proactively avoid issues.

Data analysis is key to ensuring that a system is performing as expected and helps in identifying any outstanding potential issues. Data analysis can be done manually by humans or can be programmed to be done by a system. The intent to identify the root cause once an incident has occurred is referred to as debugging and is introduced as the next key topic in this discussion regarding feedback loops.

Debugging

Debugging allows ad hoc retrospective analysis to be conducted from the information gathered through analyzing data. It helps to answer questions such as what are the other events that happened around the same time when an event occurred.

Any software service or system is bound to have unforeseen events or circumstances. These events are triggered due to an outage or loss of data or monitoring failure or the need for toil to perform a manual intervention. The active events are then responded to by either automated systems or humans. However, the response is based on the analysis of signal data that comes through the monitoring systems. These signals evaluate the impact and escalate the situation as needed and help to formulate an initial response.

Debugging is also key to an effective post-mortem technique that includes updating documentation as needed, performing root cause analysis, communicating the details of the events across teams to foster knowledge sharing, and coming up with a list of preventive actions.

The next topic in the discussion on feedback loops focuses on alerting. Alerting is essential for notifying either before an event occurs or as soon as possible after an event occurs.

Alerting

Alerting is a key follow-up to data analysis and informs the problem at hand. Real-time, or near-real-time, alerting is critical in mitigating the problem and potentially also identifying the root cause.

Alerting rules can be complex in reflecting a sophisticated business scenario and notifications can be sent when these rules are violated. Common means of notification include the following:

- An email (that indicates something happened)
- A page (that calls for immediate attention)
- A ticket (that triggers the need to address an issue sooner rather than later)

In *Chapter 2, SRE Technical Practices – Deep Dive*, we discussed the implications of setting reliability targets. For example, if the reliability target is set to 4 9's of reliability (that is, 99.99%) then this translates to 4.38 minutes of downtime in a 30-day period. This is not enough time for a human to be notified and then intervene. However, a system can be notified and can potentially take steps to remediate the issue at hand. This is accomplished through alerting.

These topics attempt to elaborate on why monitoring can be used as a feedback loop. This is critical in SRE because the goal of SRE is to maintain a balance between releasing new features and system reliability. Monitoring helps to identify issues in a timely manner (as they occur), provide an alert (when they do occur), and provide data to debug. This is key to understanding how an error budget is tracked over a period. More issues will lead to a faster burn of the error budget and thereby it becomes more important to stabilize at that point rather than release new features. However, if monitoring provides information that indicates that the current system is stable, then there will be a significant error budget remaining to prioritize new features over system stability.

Given that we have established monitoring as an essential element in providing continuous feedback to achieve continuous improvement, it is also equally important to understand the common misconceptions to avoid. This will be covered as the next topic.

Monitoring misconceptions to avoid

There are several common misconceptions when it comes to setting up monitoring for a service or system. The following is a list of such misconceptions that should be avoided:

- Monitoring should be regarded as a specialized skill that requires a technical understanding of the components involved and requires a functional understanding of the application or even the domain. This skill needs to be cultivated within the team that's responsible for maintaining the monitoring systems.

- There is no all-in-one tool to monitor a service or system. In many situations, monitoring is achieved using a combination of tools or services. For example, the latency of an API call can be monitored by tools such as *Grafana*, but the detailed breakdown of the API calls across specific methods, including the time taken for a database query, can be monitored using a tool such as *Dynatrace*.

- Monitoring shouldn't be limited to one viewpoint and instead cover multiple viewpoints. The things that matter to the end consumer might be different to what matters to the business, and may also differ from the viewpoint of the service provider.

- Monitoring is never limited to a single service. It could be extended to a set of related or unrelated services. For example, it is required to monitor a caching service as a service related to a web server. Similarly, it is important to monitor directly unrelated services, such as the machine or cluster hosting the web server.

- Monitoring doesn't always have to be complex. There may be complex business conditions that need to be checked with a combination of rules, but in many cases, monitoring should follow the **Keep it Simple Stupid** (**KISS**) principle.

- Establishing monitoring for a distributed system should focus on individual services that make up the system and should not solely focus on the holistic operation. For example, the latency of a request can be longer than expected. The focus should be on the elements or underlying services that cumulatively contribute to the request latency (which includes method calls and query responses).

- The phrase *single pane of glass* is often associated with effective monitoring, where a pane of glass metaphorically refers to a management console that collects data from multiple sources representing all possible services. But merely displaying information from multiple sources doesn't provide a holistic view of the relationships between the data or an idea of what could possibly go wrong. Instead, a single pane of glass should deliver a logical grouping of multiple services into a single workspace by establishing the correlation between monitoring signals.

Monitoring should not only focus on the symptoms, but also on their causes. Let's look at some examples:

Symptoms ("What's broken?")	Causes ("Why is it broken?")
Increased service latency	Resource usage overload
Serving HTTP 500s	Server-side errors, such as a database is down, a server is refusing new connections, and so on
Private content is "publicly" accessible	A recent software update has overwritten specific Access Control Lists, thereby making the content public
Service is returning errors to users in a specific region	Specific issues in logic are causing the errors

Essentially, the focus of monitoring should not be on collecting or displaying data, but instead *on establishing a relationship between what's broken and a potential reason for why it's broken*. There are multiple sources from which monitoring data can be collected or captured. This will be discussed as the next topic.

Monitoring sources

Monitoring data is essential in monitoring systems. There are two common sources of monitoring data. The first source is **metrics**:

- Metrics represent numerical measurements of resource usage or behavior that can be observed and collected across the system over many data points at regular time intervals. Typical time intervals for collecting metrics could be once per second, once per minute, and so on.

- Metrics can be gathered from low-level metrics provided by the operating system. In most cases, the low-level metrics are readily available as they are specific to a resource, such as database instances, virtual machines, and disks.

- Metrics can also be gathered from higher-level types of data tied to a specific component or application. In such cases, custom metrics should be created and exposed through the process of instrumentation.

- Metrics are used as input to display less granular real-time data in dashboards or trigger alerts for real-time notification.

The next source is **logs**:

- Logs represent granular information of data and are typically written in large volumes.

- Logs are not real time. Logs always have an inherent delay between when an event occurs and when it is visible in logs.

- Logs are used to find the root cause of an issue as the key data required for analysis is usually not present as a metric.

- Logs can be used to generate detailed non-time-sensitive reports using log processing systems.

- Logs can be used to create metrics by running queries against a stream of logs using batch processing systems.

Logging versus monitoring

Logging provides insights into the execution of an application. Logging can capture event records and the minutest details, along with actionable errors that could be converted into alerts. Logging essentially describes what could have happened and provides data to investigate an issue.

Monitoring, on the other hand, provides capabilities to detect issues as they happen, and alert as needed. In fact, monitoring requires logging as an essential source of information. Also, the inverse is true, that logging requires monitoring. This is because an application with fantastic logging but no monitoring is not going to help the end user.

To summarize, both metrics and logs are popular choices as monitoring sources. They are used in different situations and, in most cases, a combination of both sources is always recommended. Metrics are a good source if there are internal or external components that provide information about events and performance. Logs are best suited to track various events that an application goes through. Metrics can also be created from logs. The next topic discusses a few of the recommended monitoring strategies.

Monitoring strategies

The following are some recommended strategies while choosing a monitoring system:

- **Data should not be stale**: The speed of data retrieval becomes critical when querying vast amounts of data. If retrieval is slow, data will become stale and may be misinterpreted, with actions potentially being taken on incorrect data.

- **Dashboards should be configurable and include robust features**: A monitoring system should include interfaces that have the capabilities to display time series data in different formats, such as *heatmaps*, *histograms*, *counters*, or a *distribution*. Options to aggregate information using multiple options should be present as a configurable feature.

- **Alerts should be classified and suppressed if needed**: Monitoring systems should have the ability to set different severity levels. In addition, once an alert has been notified, it is extremely useful if there is the ability to suppress the same alert for a period that will avoid unnecessary noise that could possibly distract the on-call engineer.

These recommended strategies are implemented across two types of monitoring. This classification will be discussed as the next topic.

Monitoring types

Monitoring can be classified into the two most common types:

- Black box monitoring
- White box monitoring

Black box monitoring

Black box monitoring refers to monitoring the system based on testing externally visible behavior based on the user's perspective. This kind of monitoring does not involve any access to the technical details or build or configuration of the system. Monitoring is strictly based on testing visible behavior that is a reflection of how an end consumer would access the system. It is metaphorically referred to as a black box since the internals of the system are not opaque and there is no control or visibility in terms of what's happening inside the system. It is also referred to as server or hardware monitoring.

Black box monitoring is best used for paging incidents after the incident has occurred or is ongoing. Black box monitoring is a representation of active problems and is system-oriented, with a specific focus on system load and disk/memory/CPU utilization. Additional examples include the monitoring of network switches and network devices, such as load balancers and hypervisor level resource usage.

White box monitoring

White box monitoring is commonly referred to as application monitoring and is based on the metrics collected and exposed by the internals of the system. For example, white box monitoring can give insights into an application or endpoint performance by capturing the total number of HTTP requests or the total number of errors or the average request latency per request. In contrast, black box monitoring can only capture if the endpoint returned a successful response. White box monitoring is both symptom-and cause-oriented and this depends on how informative the internals of the system are. White box monitoring can also provide insights into future problems, as information retrieved from one internal can be the reason for an issue in another internal. White box monitoring collects information from three critical components – metrics, logs, and traces, described as follows:

- **Metrics**: These are readily available or custom created metrics that represent the state of the system in a measurable way and typically take the form of counters, gauges, and distribution.

  Metrics must be **SMART**:

  a) **S**: **Specific** (such as automation results should be at least 99% versus high quality)

  b) **M**: **Measurable** (such as results should be returned within 200 ms versus fast enough)

  c) **A**: **Achievable** (such as a service is 99.99% available versus 100% available)

  d) **R**: **Relevant** (such as observing latency versus throughput for browsing video titles)

  e) **T**: **Time Bound** (such as service is 99.99% available over 30 days versus over time)

- **Logs**: These represent a single thread of work at a single point in time. Logs reflect the application's state and are user-created at the time of application development. Logs can be structured or semi-structured, which typically includes a timestamp and a message code. Log entries are written using client libraries such as log4j and sl4j. Log processing is a reliable source of producing statistics and can also be processed in real time to produce log-based metrics.

- **Traces**: These are made up of spans. A span is the primary building block of a distributed trace that represents a specific event or a user action in a distributed system. A span represents the path of a request through one server. However, there could be multiple spans at the same time, where one span can reference another span. This allows multiple spans to be assembled into a common trace, which is essentially a visualization of requests as it traverses through a distributed system.

> **Black box monitoring versus white box monitoring – which is more critical?**
>
> Both types of monitoring are equally critical, and each is recommended based on the situation and the type of audience. Black box monitoring provides information that the Ops team typically looks at, such as disk usage, memory utilization, and CPU utilization, whereas white box monitoring provides more details on the internals of the system, which could reflect the reason for a metric produced by black box monitoring. For example, a black box metric such as high CPU utilization will indicate that there is a problem, but a white box metric such as active database connections or information on long-running queries can indicate a potential problem that is bound to happen.

To summarize, the reliability of a system can be tracked by monitoring specific metrics. However, there could potentially be multiple metrics that could be tracked and sometimes can also lead to confusion while prioritizing these metrics. The next topic lists the most important metrics to track as recommended by Google for a user-facing system. These metrics are known as the golden signals.

The golden signals

System reliability is tracked by SLOs. SLOs require SLIs or specific metrics to monitor. The types of metrics to monitor depend on the user journey tied to the service. It's strongly recommended that every service/system should measure a definite and a finite set of SLIs. So, if there is a situation where it is possible to define multiple metrics for a service, then it is recommended to prioritize the metrics to measure and monitor.

Google proposes the use of four golden signals. Golden signals refer to the most important metrics that should be measured for a user-facing system:

- **Latency**: This is an indicator of the time taken to serve a request and reflects user experience. Latency can point to emerging issues. Example metrics include *Page load/transaction/query duration*, *Time until first response*, and *Time until complete data duration*.

- **Traffic**: This is an indicator of current system demand and is also the basis for calculating infrastructure spend. Traffic is historically used for capacity planning. Example metrics include *# write/read ops, # transactions/retrievals/HTTP requests per second*, and *# active requests/connections*.

- **Errors**: This is an indicator of the rate of requests that are failing. It essentially represents the rate of errors at an individual service and for the entire system. Errors represent the rate of requests that fail explicitly or implicitly or by policy. Example metrics include *# 400/500 HTTP Codes* and *# exceptions/stack traces/dropped connections*.

- **Saturation**: This is an indicator of the overall capacity of the service. It essentially represents how full the service is and reflects degrading performance. Saturation can also indicate SLOs, resulting in the need to alert. Example metrics include *Disk/Memory quota, # memory/thread pool/cache/disk/CPU utilization,* and the *# of available connections/users on the system*.

This completes the section on monitoring, with the insights into desirable features of a monitoring system that could essentially help in creating a feedback loop, potential monitoring sources, types of monitoring, and Google's recommended golden signals, which represent the four key metrics that should be measured for a user-facing system. The next section will provide an overview of alerting and how information from the monitoring system can be used as input.

Alerting

SLIs are quantitative measurements at a given point in time and SLOs use SLIs to reflect the reliability of the system. SLIs are captured or represented in the form of metrics. Monitoring systems monitor these metrics against a specific set of policies. These policies represent the target SLOs over a period and are referred to as alerting rules.

Alerting is the process of processing the alerting rules, which track the SLOs and notify or perform certain actions when the rules are violated. In other words, alerting allows the conversion of SLOs into actionable alerts on significant events. Alerts can then be sent to an external application or a ticketing system or a person.

Common scenarios for triggering alerts include (and are not limited to) the following:

- The service or system is down.

- SLOs or SLAs are not met.

- Immediate human intervention is required to change something.

As discussed previously, SLOs represent an achievable target, and error budgets represent the acceptable level of unreliability or unavailability. SRE strongly recommends the use of alerts to track the burn rate of error budgets. If the error budget burn rate is too fast, setting up alerts before the entire budget is exhausted can work as a warning signal, allowing teams to shift their focus on system reliability rather than push risky features.

The core concept behind alerting is to track events. The events are processed through a time series. **Time series** is defined as a series of event data points broken into successive equally spaced windows of time. It is possible to configure the duration of each window and the math applied to the member data points inside each window. Sometimes, it is important to summarize events to prevent false positives and this can be done through time series. Eventually, error rates can be continuously calculated, monitored against set targets, and alerts can be triggered at the right time.

Alerting strategy – key attributes

The key to configuring alert(s) for a service or a system is to design an effective alerting strategy. To measure the accuracy or effectiveness of a particular alerting strategy, the following key attributes should be considered during the design.

Precision

From an effectiveness standpoint, alerts should be bucketized into relevant alerts and irrelevant alerts. **Precision** is defined as the proportion of events detected that are significant.

The following is a mathematical formula for calculating precision:

$$Precision\ (in\ \%) = \frac{Relevant\ Alerts}{Relevant\ Alerts + Irrelevant\ Alerts} * 100$$

In other words, if precision needs to be 100%, then the count of irrelevant alerts should be 0. Precision is a measure of exactness and is often adversely affected by false positives or false alerts. This is a situation that could occur during a low-traffic period.

Recall

An alert needs to capture every significant event. This means that there should not be any missed alerts. **Recall** is defined as the proportion of significant events detected.

The following is a mathematical formula for calculating recall:

$$Recall\ (in\ \%) = \frac{Relevant\ Alerts}{Relevant\ Alerts + Missed\ Alerts} * 100$$

In other words, if recall needs to be 100%, then every significant event should result in an alert and there should not be any missed alerts. Recall is a measure of completeness and is often adversely affected by missed alerts.

Detection time

Detection time is defined as the time taken by a system to notice an alert condition. It is also referred to as the time to detect. Long detection times can negatively impact the error budget. So, it is critical to notify or raise an alert as soon as an issue occurs. However, raising alerts too fast will result in false positives, which will eventually lead to poor precision.

Reset time

Reset time is defined as the length or duration of time the alerts are fired after an issue has been resolved. Longer reset times will have an adverse impact because alerts will be fired on repaired systems, leading to confusion.

This concludes an introduction to key attributes that are critical in defining an effective alerting strategy. The next topic elaborates on a potential approach to define an effective alerting strategy.

Alerting strategy – potential approaches

SRE recommends six different approaches to configure alerts on significant events. Each of these approaches addresses a different problem. These approaches also offer a certain level of balance across key attributes, such as precision, recall, detection time, and reset time. Some of these approaches could solve multiple problems at the same time.

Approach # 1 – Target error rate >= Error budget (with a shorter alert window)

In this approach, a shorter alert window or smaller window length is chosen (for example, 10 minutes). Smaller windows tend to yield faster alert detection and shorter reset times but also tend to decrease precision because they tend toward false positives or false alerts.

As per this approach, an alert should be triggered if the target error rate equals or exceeds the error budget within the defined shorter alert window.

Consider an example with the following input parameters:

- The expected SLO is 99.9% over 30 days, resulting in 0.1% as the error budget.

- Alert window to examine: 10 minutes.

Accordingly, a potential alert definition would be the following:

If the SLO is 99.9% over 30 days, alert if the error rate over the last 10 minutes is >= 0.1%.

In approach # 1, a shorter time window results in alerts being fired more frequently, but tends to decrease precision.

Approach # 2 – Target error rate >= Error budget (with a longer alert window)

In this approach, a longer alert window or larger window length is chosen (for example, 36 hours). Larger windows tend to yield better precision, but will have longer reset and detection times. This means that the portion of the error budget spent before the issue is detected is also high.

As per this approach, an alert should be triggered if the target error rate equals or exceeds the error budget within the defined larger time window.

Consider an example with the following input parameters:

- **The expected SLO is 99.9% over 30 days**: Resulting in 0.1% as the error budget

- **Alert window to examine**: 36 hours

Accordingly, a potential alert definition would be the following:

If the SLO is 99.9% over 30 days, alert if the error rate over the last 36 hours is >= 0.1%.

In approach # 2, a longer time window results in higher precision, but the alert will be fired less frequently and may result in a higher detection time.

Approach # 3 – Adding a duration for better precision

In this approach, a duration parameter can be added to the alert criteria, so that the alert won't fire unless the value remains above the threshold for that duration. The choice of threshold also becomes significant as some of the alerts can go undetected if the threshold is too high. For example, if the duration window is 10 minutes and the signal data was up for 9 minutes but returned to normal before the 10th minute, the error budget will be consumed, but alerts will go undetected or will not get fired. With the right selection of duration parameter and threshold, this approach enables the error to be spotted quickly, but treats the error as an anomaly until the duration is reached:

- The advantage is that the alert fired after a defined duration will generally correspond to a significant event, and so will increase precision.
- The disadvantage is that the error will continue to happen for a larger window and, as a result, will lead to a deteriorating recall.

A longer alert window or larger window length is recommended (for example, 36 hours). Larger windows tend to yield better precision, but will have longer reset and detection times.

As per this approach, an alert should be triggered if the target error rate equals or exceeds the error budget within the defined larger time window.

Consider an example with the following input parameters:

- **The expected SLO is 99.9% over 30 days**: Resulting in 0.1% as the error budget
- **Alert window to examine**: 36 hours

Accordingly, a potential alert definition would be as follows:

If the SLO is 99.9% over 30 days, alert if the error rate over the last 36 hours is >= 0.1%.

In approach # 3, a longer duration window also means that the portion of the error budget spent before the issue is detected is high, but is highly likely to indicate to a significant event.

Approach # 4 – Alert regarding the burn rate

In this approach, alerts should be defined based on burn rate. **Burn rate** is defined as how fast, relative to the SLO, the service consumes the error budget.

For example, with a 0.1% error budget over 30 days, if the error rate is constant at 0.1% across the 30 days, then the budget is spent equally and 0 budget remains at the end of the 30th day. In this case, the burn rate is calculated as 1. But if the error rate is 0.2%, then the time to exhaustion will be 15 days, and the burn rate will be 2.

When alerting based on burn rate, the following are two possible alerting policies:

- **Fast burn alert**: An alert is fired because of a sudden large change in consumption of the error budget. If not notified, then the error budget will exhaust quicker than normal. For a fast burn alert, the lookback period should be shorter (say 1 to 2 hours), but the threshold for the rate of consumption for the alert should be much higher (say 10x times) than the ideal baseline for the lookback period.

- **Slow burn alert**: An alert is not fired until a condition is continuously violated for a long period. This kind of alert only consumes a small percentage of the error budget when it occurs and is significantly less urgent than a fast burn alert. For a slow burn alert, the lookback period should be longer (say 24 hours) but the threshold is slightly higher (say 2x times) than the baseline for the lookback period.

So, as per approach # 4, alerts will be fired if the burn rate is greater than the desired burn rate at any point in time. This approach provides better precision over a shorter time window with good detection times.

Approach # 5 – Multiple burn rate alerts

The same alerting strategy doesn't need to always work for a service. This might depend on factors such as the amount of traffic, the variable error budget, or peak and slow periods. For example, during a holiday shopping season, it is common that the SLO for a service will be higher than normal. This means a lower error budget during peak season and this will revert to a slightly less strict error budget during off-peak seasons.

In this approach, instead of defining a single condition in an alerting policy, multiple conditions can be defined to get better precession, recall, detection time, and reset time. Each condition can have a different level of severity and a different notification channel to notify the alert based on the nature of severity.

For example, an alerting policy can be defined with the following multiple conditions:

- *Trigger an alert if 10% of the budget is consumed over a time duration of 3 days and notify by creating a ticket.*

- *Trigger an alert if 5% of the budget is consumed over a time duration of 6 hours and notify through a page.*

- *Trigger an alert if 2% of the budget is consumed over a time duration of 1 hour and notify through a page.*

In approach # 5, multiple conditions can be defined for a single alerting policy. Each condition could result in a different action that could potentially represent the severity level for the alert.

Approach # 6 – Multiple burn rate alerts across multiple windows

This is an extension of approach # 5, where the major difference is to use multiple windows to check whether the error rate exceeds the error budget rather than a single window for a single condition. This will ensure that the alert raised is always significant and is actively burning the error budget when it gets notified.

This helps to create an alerting framework that is flexible and shows the direct impact based on the severity of the incident. A flexible window emphasizes or confirms whether the alert condition is active in the last specified duration. This helps to immediately troubleshoot when alerted. The flip side is that such conditions also involve multiple variables that might add to maintenance in the long term.

For example, an alerting policy can be defined with the following multiple conditions:

- *Trigger an alert if 10% of the budget is consumed over a time duration of 3 days and is currently being consumed over the last 6 hours. Notify by creating a ticket.*

- *Trigger an alert if 5% of the budget is consumed over a time duration of 6 hours and is currently being consumed over the last 30 minutes. Notify through a page.*

- *Trigger an alert if 2% of the budget is consumed over a time duration of 1 hour and is currently being consumed over the last 5 minutes. Notify through a page.*

Approach # 6 is potentially an extension of approach # 5, where multiple conditions can be defined for an alerting policy where each condition, if breached, can result in a different action or notification. The difference is that approach # 6 emphasizes specifying an alert window that could confirm that the fired alert is potentially active.

These approaches are well suited for many situations that could require an alerting strategy. However, there could be a specific situation where the traffic received by the service is less or low. The next topic discusses approaches on how to deal with such situations.

Handling service with low traffic

If a service receives a smaller amount of traffic, a single or a smaller number of failed requests might result in a higher error rate, indicating a significant burn of the error budget. SRE recommends a few options for handling a low-traffic service:

- **Generating artificial traffic**: This option provides more signals or requests to work with. However, a significant amount of engineering effort is required to ensure that the artificial traffic behavior matches real user behavior closely.

- **Combining services**: This option recommends bucketizing similar low-request services into a single group representing a single function. This will result in higher precision and fewer false positives. However, careful consideration needs to be given when combining the services into a single group. This is because the failure of an individual service might not always result in the failure of the overall function and, as a result, will not result in an alert for a significant event.

- **Modifying clients**: This option is used to deal with ephemeral failures, especially if it is impractical to generate artificial traffic or combine services into a group. The impact of a single failed request can be reduced by modifying the client and implement exponential backoff. Additionally, fallback paths should be set up to capture the request for eventual execution post-backoff.

- **Lowering the SLO or increasing the window**: This option is the simplest way to handle low-traffic services. This will reduce the impact of a single failure on the error budget. However, lowering the SLO is also a way to lower the expectations on how the service should behave.

Given that we learned about topics specific to creating an effective alerting strategy, the next logical step is to learn about steps required to establish an SLO altering policy.

Steps to establish an SLO alerting policy

The following is the sequence of steps to establish an SLO alert policy:

1. **Select the SLO to monitor**: Choose the suitable SLO for the service. It is recommended to monitor only one SLO at a time.

2. **Construct an appropriate condition**: It is possible to have multiple conditions for an alerting policy where the condition is different for a slow burn when compared to a fast burn.

3. **Identify the notification channel**: Multiple notification channels can be selected at the same time for an alerting policy.

4. **Include documentation**: It is recommended to include documentation about the alert that might help to resolve the potential underlying issue.

5. **Create an alerting policy**: The policy should be created in the monitoring system of choice, either through a configuration file or CLI, or through the console (based on the features supported by the monitoring system).

This concludes establishing the blueprint for the SLO alerting policy. The next topic introduces the desirable characteristics of an alerting system.

Alerting system – desirable characteristics

The alerting system should control the number of alerts that the on-call engineers receive. The following is a shortlist of desirable characteristics that an alerting system should possess:

- Inhibit certain alerts when others are active.

- Remove or silence duplicate alerts from multiple sources that have the same label sets.

- Fan-in or fan-out alerts based on their label sets when multiple alerts with similar label sets fire.

This completes the section on alerting with insights into constructs that are required to define an effective alerting policy that includes possible approaches and key attributes, such as precision, recall, detection time, and reset time. The next section will provide an overview of time series, their structure, cardinality, and metric types.

Time series

Time series data is the data that collectively represents how a system's behavior changes over time. Essentially, applications relay a form of data that measures how things change over time. Time is not only regarded as a variable being captured; time is the primary focal point. Real-world examples of time series data include the following:

- Self-driving cars that continuously collect data to capture the ever-changing driving conditions or environment

- Smart homes that capture events such as a change in temperature or motion

---

Metric versus events

Metrics are time series measurements gathered at regular intervals. **Events** are time series measurements gathered at irregular time intervals.

The following are some characteristics that qualify data as time series data:

- Data that arrives is always recorded as a new entry.

- Data arrives in time order.

- Time is the primary axis.

> **Adding a time field to the dataset is not the same as time series data**
>
> Data related to a sensor is being collected in a non-time series database. If the sensor collects a new set of data, then writing this data will overwrite the previously stored data in the database with updated time. The database will eventually return the latest reading, but will not be able to track the change of data over a period of time. So, a non-time series database tracks changes to the system as **UPDATES**, but a time series database tracks changes to the system as **INSERTS**.

The next topic discusses the structure of time series.

Time series structure

Monitoring data is stored in time series. Each individual time series data has three pieces of information (refer to *Figure 3.1*):

- **Points**: Refers to a series of (timestamp, value) pairs. The value is the measurement, and the timestamp is the time at which the measurement was taken.

- **Metric**: Refers to the name of the metric type that indicates how to interpret the data points. This also includes a combination of values for the metric labels.

- **Monitored resource**: Refers to the monitored resource that is the source of the time series data, and one combination of values for the resource's label.

The following screenshot shows the structure of time series data:

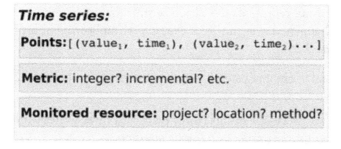

Figure 3.1 – Structure of time series data

Let's look at an example:

The following screenshot shows an illustration of how time series data is represented for metric types with sample values: `storage.googleapis.com/api/request_count`:

```
[bucket: 1234,    response_code: OK,    method: read] {(3, Wed 2:00pm),
                                                       (2, Wed 2:05pm),
                                                       (8, Wed 2:10pm),
                                                       ...}
[bucket: 1234,    response_code: OK,    method: write]{(1, Wed 2:01pm),
                                                       (2, Wed 2:04pm),
                                                       (7, Wed 2:09pm),
                                                       ...}
[bucket: 1234,    response_code: FAIL, method: write]{(1, Wed 2:01pm),
                                                       (0, Wed 2:04pm),
                                                       (0, Wed 2:09pm),
                                                       ...}
[bucket: 9876,    response_code: OK,    method: read] {(2, Wed 1:59pm),
                                                       (4, Wed 2:05pm),
                                                       (3, Wed 2:10pm),
                                                       ...}

...
     monitored                     metric                      metric
  resource label                   labels                       data
```

Figure 3.2 – Illustration of a GCP Cloud storage metric with sample values

The preceding screenshot represents the various events that were captured as part of the metrics as they happened over time. For example, the following information can be extracted from *Figure 3.2*:

- There are a total of eight registered events in bucket `1234` between `2:00` pm and `2:10` pm.

- This includes six successful events (three reads and three writes with different timestamps) and three unsuccessful write attempts.

The next topic introduces the concept of cardinality with respect to time series data.

Time series cardinality

Each time series is associated with a specific pair of metric and monitored resource types, but each pair can have many time series. The possible number of time series is determined by the cardinality of the pair: the number of labels and the number of values each label can take on.

For example, a time series metric is represented as a combination of two labels: **zone** and **color**. There are two zones (east and west) and three colors for each zone (red, green, and blue). So, the cardinality for this metric is six. The following is the potential time series data:

```
Request_by_zone_and_color{zone="east",color="red"}
Request_by_zone_and_color{zone="east",color="green"}
Request_by_zone_and_color{zone="east",color="blue"}
Request_by_zone_and_color{zone="west",color="red"}
Request_by_zone_and_color{zone="west",color="green"}
Request_by_zone_and_color{zone="west",color="blue"}
```

With metric cardinality, there are a couple key pointers to bear in mind:

- Metric cardinality is a critical factor with respect to performance. If the cardinality is too high, this means that there is a lot of time series data, resulting in higher query response times.
- For custom metrics, the maximum number of labels that can be defined in a metric type is 10.

The next topic introduces the common metric types with respect to time series data.

Time series data – metric types

Each time series includes a metric type to represent its data points. The metric type defines how to interpret the values relative to one another. The three most common types of metric are as follows:

- **Counter**: A counter is a cumulative metric that represents a monotonically increasing function whose value can only increase (but can never decrease) or be reset to zero on restart; for example, a counter to represent the number of requests served, tasks completed, or errors observed up to a particular point in time.
- **Gauge**: A gauge metric represents a single numerical value that can arbitrarily go up and down. It is useful for monitoring things with upper bounds.

Examples of a gauge include the size of a collection or map or the number of threads in a running state. Gauges are also typically used for measured values such as temperatures or current memory usage, but also *counts* that can go up and down, such as the number of concurrent requests.

- **Distribution**: A distribution metric is used to track the distribution of events across configurable buckets; for example, measure the payload sizes of requests hitting the server.

This completes the section on time series, where key concepts related to time series structure, cardinality, and possible metric types were summarized. This also brings us to the end of the chapter.

Summary

In this chapter, we discussed the concepts related to monitoring, alerting, and time series that are critical in tracking SRE technical practices, such as the SLO and error budgets. We also discussed the differences between black box monitoring and white box monitoring. In addition, we examined the four golden signals as recommended by Google to be the desired SLI metrics for a user-facing system.

In the next chapter, we will focus on the constructs required to build an SRE team and apply cultural practices such as handling facets of incident management, being on-call, avoiding psychological safety, promoting communication and collaboration, and knowledge sharing.

Points to remember

The following are some important points to remember:

- Black box monitoring is based on testing externally visible behavior.
- White box monitoring is based on the metrics collected and exposed by the internals of the system.
- Metrics must be specific, measurable, achievable, relevant, and time-bound.
- The four golden signals recommended for a user-facing system are latency, traffic, errors, and saturation.
- Latency can point to emerging issues, and traffic is historically used for capacity planning.

- Errors represent the rate of requests that fail explicitly or implicitly or by policy.

- Saturation represents how full the service is and reflects degrading performance.

- Precision is defined as the proportion of events detected that are significant.

- Recall is defined as the proportion of significant events detected.

- Detection time is defined as the time taken by a system to notice an alert condition.

- Reset time is defined as the length or duration of time the alerts are fired after an issue has been resolved.

- An individual time series data has three critical pieces of information – points, metrics, and monitored resources.

Further reading

For more information on GCP's approach to DevOps, read the following articles:

- **SRE**: `https://landing.google.com/sre/`

- **SRE Fundamentals**: `https://cloud.google.com/blog/products/gcp/sre-fundamentals-slis-slas-and-slos`

- **SRE Youtube Playlist**: `https://www.youtube.com/watch?v=uTEL8Ff1Zvk&list=PLIivdWyY5sqJrKl7D2u-gmis8h9K66qoj`

- **Metrics, time series, and resources**: `https://cloud.google.com/monitoring/api/v3/metrics`

Practice test

Answer the following questions:

1. Select the monitoring option that works based on the metrics exposed by the internals of the system.

    a) Alert-based monitoring

    b) White box monitoring

    c) Log-based monitoring

    d) Black box monitoring

2. Select the monitoring source that doesn't provide information in near-real time.

    a) Logs

    b) Metrics

    c) Both

    d) None of the above

3. From the perspective of a fast burn alerting policy, select the appropriate threshold in relative comparison to the baseline for the defined lookback interval.

    a) The threshold = the baseline.

    b) The threshold is < the baseline.

    c) The threshold is significantly higher than the baseline.

    d) The threshold is slightly higher than the baseline.

4. Select the appropriate options for sending alerts

    a) Email

    b) Page

    c) Text

    d) All of the above

5. Select the monitoring that is best suited to paging incidents.

    a) Alert-based monitoring

    b) White box monitoring

    c) Log-based monitoring

    d) Black box monitoring

6. Which of the following is not part of Google's recommended golden signals?

    a) Traffic

    b) Throughput

    c) Saturation

    d) Errors

7. Which of the following alerting policies recommends a longer lookback window?

    a) Fast burn

    b) Slow burn

    c) Both

    d) None

8. Which of the following represents the action of collecting, processing, aggregating, and displaying real-time quantitative data relating to a system, such as query counts and types, error counts and types, processing times, and server lifetimes?

    a) Alerting

    b) Monitoring

    c) Debugging

    d) Troubleshooting

9. Which of the following represents time series measurements gathered over irregular time intervals?

    a) Metrics

    b) Events

    c) Logs

    d) Trace

10. Which of the following is not a suitable source of white box monitoring?

    a) Metrics

    b) Load balancer

    c) Logs

    d) Traces

Answers

1. (b) White box monitoring.

2. (a) Logs.

3. (c) The threshold is significantly higher than the baseline. The recommended level is 10x.

4. (d) All of the above, including email, pages, and texts.

5. (d) Black box monitoring.

6. (b) Throughput. The four golden signals are latency, errors, traffic, and saturation

7. (b) Slow burn alert policy

8. (b) Monitoring

9. (b) Events. Events are time series measurements gathered at irregular time intervals. Metrics are time series measurements gathered at regular time intervals.

10. (b) Load balancer. It is best suited as a source for black box monitoring. White box monitoring collects information from three critical components: metrics, logs, and traces.

4
Building SRE Teams and Applying Cultural Practices

The last three chapters introduced the fundamentals of **Site Reliability Engineering (SRE)**, traced its origins, laid out how SRE is different than DevOps, introduced SRE jargon along with its key technical practices such as **Service Level Agreements (SLAs)**, **Service Level Objectives (SLOs)**, **Service Level Indicators (SLIs)**, and **Error Budgets**, and focused on monitoring and alerting concepts to target reliability.

This chapter will focus on the fundamentals required to build SRE teams and apply cultural practices such as handling facets of incident management, being on call, achieving psychological safety, promoting communication, collaboration and knowledge sharing. These fundamentals and cultural practices can be used as a blueprint for teams or organizations that want to start their SRE journey.

In this chapter, we're going to cover the following main topics:

- **Building SRE teams** – Staffing, creating SRE teams, and engaging the team
- **Incident management** – Incident life cycle and constructs to handle the incident
- **Being on call** – Challenges to tackle, operational overload, and effective troubleshooting
- **Psychological safety and fostering collaboration** – Factors to achieve psychological safety, unified vision, communication, and collaboration and knowledge sharing

Building SRE teams

Google defined the principles of SRE by applying the concepts of software engineering to system operations. Google was implementing these principles even before the term *DevOps* was coined. They developed best practices over a period of time and essentially considered SRE as their secret sauce for efficient running of their products. With the advent of **Google Cloud Platform (GCP)**, Google became more vocal about the SRE principles and their relevance for the success of their customers that deal with maintaining, running, and operating distributed systems on GCP.

Given SRE is a prescriptive way of doing DevOps, more and more organizations (and this also includes non-GCP customers) are currently tending toward implementing the principles of SRE in a quest to find a balance between service reliability and development velocity. Such organizations will face the following challenges:

- How do you staff an SRE team?
- How do you implement or run an SRE team?
- When and how often are SRE teams engaged during the life cycle of a service?

The following sub-sections answer these questions based on the best recommendations from Google.

Staffing SRE engineers (SREs)

An SRE team consists of SRE engineers or SREs. At the outset, SREs also run operations. It is hard to find seasoned SREs. However, one way to build an SRE team is to hire system administrators who have worked on operations along with having experience in scripting/coding. These personnel can be further trained with software engineering skills.

The following is a list of recommended skills that personnel hired as SRE engineers or SREs should possess or ultimately be trained on:

- **Operations and software engineering**: SREs should have experience of running a production system and have an understanding of the software or application that needs to be supported.

- **Monitoring systems**: SLOs are key to maintaining the reliability of a service and SREs need to understand the functioning of monitoring systems in order to track SLOs and their connected constructs such as SLIs and Error Budgets.

- **Production automation**: SREs can effectively scale operations by ensuring that the same task is not performed manually and instead automation is put in place. This requires an understanding of how to automate the process.

- **System architecture**: SREs can effectively scale an application or service, and this requires a deep understanding of the system architecture.

- **Troubleshooting**: SREs are regularly required to be on call to solve problems of different types and complexities. This requires an inquisitive and analytical approach to solving the problems in hand.

- **Culture of trust**: SREs and developers share the ownership of a service through the concept of Error Budgets. This requires SREs to build a trust with the development team through effective communication.

- **Incident management**: Failures or issues are inevitable and one of the key functions of SREs is to handle an incident. This requires the ability to technically troubleshoot and establish a communication framework specific for managing incidents.

In addition to the recommended skills, SRE engineers or SREs should also possess certain character traits such as resilience, assertiveness, and flexibility. These characteristics will help them to deal with difficult situations and use reasoning in case of ambiguous situations, and strike a balance between development velocity and reliability. The next topic will deep-dive into the types of SRE team implementations.

SRE team implementations – procedure and strategy

There are six recommended SRE team implementations:

- Kitchen sink/everything SRE team

- Infrastructure SRE team

- Tools SRE team

- Product/application SRE team

- Embedded SRE team
- Consulting SRE team

Each implementation type suits a specific organization based upon the complexity of services that the organization provides, the size of the organization, the number of kinds of development teams in terms of varying scopes, and the organization's level of adoption with respect to applying SRE principles. Some organizations can also implement more than one type based on their needs. Let's look at the different team implementations in detail.

Kitchen sink/everything SRE team

The **kitchen sink/everything SRE team** is best suited for organizations that want to start their SRE journey. In most cases, organizations prefer to start with a single SRE team. The scope of such a team is unbounded. The kitchen sink/everything SRE team is recommended for organizations that currently have a limited scope because there are fewer applications, resulting in fewer user journeys.

Given that there is only a single SRE team, this type of team can act as a binding factor between development teams and can also spot patterns across projects to provide effective solutions in case of incidents. At the same time, identifying engineers for SRE teams with the required skill sets can be quite challenging. In addition, with the possibility of constantly changing scope, the kitchen sink team implementation can suffer from lack of clarity in terms of well-defined team goals and can eventually lead to operational overload.

Infrastructure SRE team

The **infrastructure SRE team** is best suited for organizations that have multiple development teams with varying scopes, complexities, and infrastructure requirements. The core focus of an infrastructure team is on behind-the-scenes tasks that help the development teams to get their job done easier and faster. An infrastructure SRE team is less focused on customer-facing code and more on maintaining the shared services and components related to stabilizing the infrastructure.

The main advantage of an infrastructure SRE team is to provide the same highly reliable infrastructure to the development teams to simulate the behavior that could potentially happen in a production environment. In addition, the infrastructure team also establishes the structure to access such environments using the principles of least privilege and adhering to the required compliance standards.

On the flip side, given that the infrastructure SRE team doesn't directly work with customers and instead works with internal teams, there is always a tendency to over-engineer while defining standards for the development teams. Additionally, their usage behavior might also differ in comparison to the real world.

At times, there might be multiple infrastructure teams based on the size of the company or its product lines. This could result in duplicate effort with respect to the manner in which infrastructure will be provisioned. This is something that needs to be closely monitored and can be potentially avoided through cross-infrastructure team meet-ups and regular updates.

Tools SRE team

The **tools SRE team** is best suited for organizations with multiple development teams, where the development teams need a standard way to build or implement software that can measure, maintain, and improve the system reliability of a service, for example implementing a common framework that could allow the development team to implement custom metric instrumentation in their services, which could eventually help to measure reliability of a service.

The tools SRE team aims to build highly specialized reliability-related tooling and ultimately define production standards. One of the major disadvantages is to clearly draw the line between a tools SRE team and an infrastructure SRE team, as they can be perceived as being very similar because they are aimed at providing focused help to development teams but at their core, their focus areas are different. The tools SRE team focuses more on improving the reliability of the individual service whereas the infrastructure SRE team focuses more on improving the reliability of the supporting infrastructure that runs the service.

Product/application SRE team

The **product/application SRE team** is best suited for organizations that already have a kitchen sink, infrastructure, or tools SRE team and in addition have a critical user-facing application with multiple key services and high reliability needs for each service. The need for multiple reliable services requires a dedicated set of SRE teams focused on the product. This approach helps in setting out a clear charter for the team and also directly relates to business priorities.

On the flip side, as the services or products grow, there will be a constant need to add more SRE teams. This also increases the chance that there might be duplication of effort among product/application SRE teams.

Embedded SRE team

The **embedded SRE team** is best suited for organizations where a dedicated SRE team is required only for a specific time period or for an implementation focused on specific functions. In this SRE setup, the SRE engineers are actually embedded with the development team and this engagement is either scope bound, or time bound.

SRE engineers are hands-on on a day-to-day basis with respect to handling code and the required configuration. An embedded SRE team can also effectively be used to drive adoption with respect to the proposals or changes put forward by the infrastructure or tools SRE team. The major advantage is to build SRE expertise for specific focus areas. However, on the flip side, given each SRE team will work with a different development team, there might be a lack of standard practices.

Consulting SRE team

The **consulting SRE team** is best suited for organizations where the complexity is significant and the existing SRE teams cannot support the ever-increasing demands of their customers. Consulting SRE teams are similar to embedded SRE teams but are focused on the end customers and typically move away from changing any of the customer code or configuration. They could, however, write code to build internal tools for themselves or for their developer counterparts, similar to a tools SRE team.

Consulting SRE teams are a good fit if additional support is required on a temporary basis for existing SRE teams but, on the flip side, the SREs might not have the required context to make a balanced decision.

This completes the section on the various types of SRE team implementations. Essentially, there is no specific SRE team implementation that is recommended. It's common for an organization to implement a combination of these teams and the decision depends on multiple factors such as the size of the organization, the maturity level of adopting SRE, and their current focus.

To summarize, if the intent of an organization is to get started with SRE, then kitchen sink implementation is probably the best place to start. If the size of the development organization is small but they have a critical user-facing application, it is possible to move from a kitchen sink implementation to a product/application SRE team implementation.

However, if there are many development teams with different infrastructure needs, then it is possible to implement an infrastructure SRE team from a kitchen sink one. A tools SRE team is best suited for organizations that try to include common frameworks to bring in standardization in their software development process. A tools SRE team can complement an infrastructure SRE team.

Embedded SRE teams are more laser-focused for a specific initiative where the development and SRE team work together to reliably implement a service. This can be seen in both small-sized and large-sized teams, based on the criticality of the initiative. Finally, a consulting team is best suited for complex organizations that already have implemented other possible SRE teams but still require a consulting team to support the increasing needs of a rapidly expanding customer base.

The next topic will elaborate on how SRE engineers are engaged throughout the life cycle of a service or an application.

SRE engagement model

The core principles of SRE are focused on maximizing development velocity while maintaining reliability. As the service goes through its life cycle phases, SRE team(s) can continuously contribute for the betterment of the service and these teams can be engaged at different stages (described as follows) with varying capacities.

Architecture and design

Let's explore what this phase entails for SREs (SRE engineers):

- SREs' engagement level is very deep (and high) during this phase as SRE teams can bring in their production expertise, offer various insights, and effectively can co-design the service.

- SREs validate the design and architecture by probing the design choices and validating any assumptions that were taken. This avoids any potential re-design.

- SREs put forward best practices during the design phase, such as resilience, by identifying the single points of failure.

- SREs recommend the best infrastructure systems based on prior experience and potential resource footprint in comparison with predicted load.

- SREs identify effective user journeys as part of the design process.

The next phase is active development.

Active development

Let's explore what this phase entails for SREs (SRE engineers):

- Once the design is in place, SREs (SRE engineers) can help development teams ensure that the service is developed with production in mind. This includes capacity planning, identifying load requirements, and adding resources for redundancy.

- SREs help to plan for spikes and overloads. One way to handle the same is to use load balancers and also evaluate the capacity that they need to be configured. Setting up load balancers early in the development environments is a good start where the behavior can be initially assessed and hardened more during performance testing that simulates a production setup.

- SREs think through the identified user journeys and work with development teams (along with other stakeholders) to come up with approaches to meet their SLI and SLO requirements.

- SREs engage to add observability that includes configuring monitoring, alerting, and performance tuning. This helps in setting up the required infrastructure to track SLIs and SLOs.

After development is complete, the service is made available to a limited number of users in the following phase.

Limited availability

Let's explore what the limited availability phase entails for SREs (SRE engineers):

- Limited availability refers to the alpha (a release that is partially complete or still in development) and beta (a release where service is available to limited users for the first time post development) releases prior to general availability. During limited availability, the number of users, potential use cases, complexity, and performance requirements change significantly when compared to the development phase.

- SREs focus on measuring the performance based on the changing demands and essentially evaluate reliability by defining SLOs.

- SREs involve development teams in establishing operational practices similar to a real-time production environment. This helps to simulate the experience for internal teams in terms of what to expect when in production.

- SREs establish incident response teams by assigning specialized roles, conducting mock incident drills, and getting ready for a real-world scenario.

The next phase is general availability. This is the phase where the service has reached a stable state and is available for use by a wider audience or by other services.

General availability

Let's explore what the general availability phase entails for SREs (SRE engineers):

- A service moves to general availability phase only if it passes the **Production Readiness Review** (**PRR**). This is potentially the longest phase that SREs are involved with and also own.

- SREs perform the majority of the operational work and own the incident response, with some help from the development team.

- It is possible that development team members work with SREs on a rotational basis so that the development team will have insights on operational load.

- SREs will focus on tracking the operational load and accompanying SLOs for the service. This ensures that Error Budgets are not exhausted, and new features can be rolled out.

When the current version of the service is going to be replaced, it enters the next phase, called depreciation.

Depreciation

Let's explore what the depreciation phase entails for SREs (SRE engineers):

- This refers to the phase when a new version of the system will soon come into play and replace the current version. So, new users or improved features are not added anymore. The focus shifts to transitioning users from existing to new systems through engineering means.

- SREs continue to support the existing system till end of life and work in parallel with development team(s) on new services by circling back to the architecture and design phase.

The last phase is the abandoned phase. This phase explains what happens to the service once the depreciation phase has passed.

Abandoned

Let's explore what the abandoned phase entails for SREs (SRE engineers):

- Once the service end of life has been reached, the service is abandoned and SREs' engagement with respect to the service ends. A service reaches end of life either if the service is no longer supported by the development team or the service is not required by customers anymore.

- Development teams resume operational support and SREs support service incidents on a best-effort basis.

This completes the section on the SRE engagement model with the emphasis on how SRE engineers have a significant impact across the life cycle of the service right from its conceptualization to its end of life. The next section details a key SRE cultural practice called incident management.

Incident management

Incident management is one of the key roles of an SRE engineer. An incident is defined as an event that indicates the possibility of an issue with respect to a service or an application. The nature of the issue can be minor in nature in the best case or, in contrast, can be an outage in the worst case. An incident can be triggered by an alert that was set up as part of monitoring the service or application.

An alert is an indication that SLO objectives with respect to the service are being violated or are on track to be violated. Sometimes, and specifically for an external-facing application, an incident can be triggered by an end user complaining via social media platforms. Such incidents include an additional layer of retrospection on how or why the current alerting system put in place failed to identify the incident.

Effective incident management is a critical SRE cultural practice that is key to limiting the disruption caused by an incident and is critical to resuming normal business operations as early as possible. There are two main parts to incident management:

- The technical troubleshooting aspect (where the emphasis is on either mitigating the issue or resolving the issue)

- The effective communication aspect (where the emphasis is on ensuring the right folks are involved in the right roles, and stakeholders with respect to the service consumers are informed in a timely manner)

Organizations approach incident management in different ways but, if not approached in the right manner, it will result in an unmanaged incident. The following are two main characteristics of an unmanaged incident:

- **Emphasis on the technical problem**: An operational issue typically has a cascading effect, especially in a complex system. So, it is common to miss the big picture while attempting to solve the technical issue at hand from a single viewpoint.

- **Poor communication**: In an attempt to solve the incident at hand, it is possible that not enough attention is given to the communication aspect. This will have both internal and external implications. Internally, lack of communication will lead to an inefficient use of resources. Sometimes it can also lead to ad hoc involvement, where multiple people can work on the same problem that could also result in multiple people making changes to the system at the same time (which is not ideal). Externally, lack of communication will lead to frustrated customers, which will result in a loss of trust toward the service or application provider.

The next set of topics in this section detail the concepts required to effectively implement incident management, starting with the topic on the incident life cycle.

Incident life cycle

Prior to coming up with an effective incident management strategy, it is important to understand the life cycle of an incident. The following is a state diagram that provides insights into the various states of an incident:

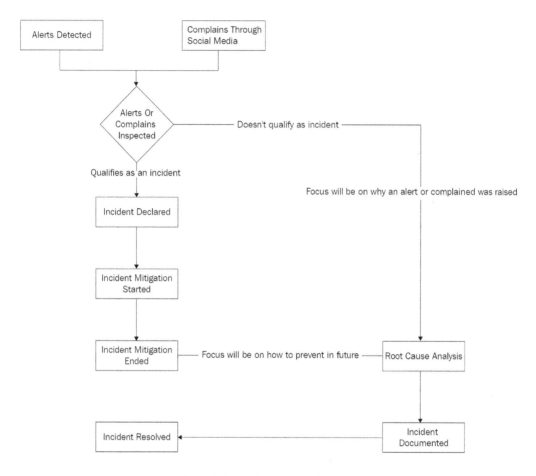

Figure 4.1 – Life cycle of an incident

To summarize, based on configured alerts or complaints from social media, an incident is identified. The identified incident is validated to check whether it qualifies. If it qualifies, incident mitigation is initiated. After incident mitigation, a root cause analysis is initiated for the incident. The details of the incident are documented, a ticket is created, and after that the incident will move to resolved status. The next topic goes into the important concepts related to effective incident management.

Elements of effective incident management

A structured incident response comprising of multiple elements is critical for effective incident management. These elements allow us to focus on core SRE cultural practices such as communication, collaboration, learning, knowledge sharing, organized flow of information, and effective customer notifications. This helps the team to deal with different situations and respond as quickly as possible. A structured incident response helps in reducing duplication of effort and creates visibility of any individual's activity within the team.

The following are some of the critical elements for effective incident management.

Declaring an incident

In the case of an event leading to an incident, SREs or operators are often in a dilemma when deciding when to declare that incident. If they wait for a longer duration to declare an incident, then it might be too difficult to mitigate the negative impact that the incident can cause to the end users. However, if they declare it too early, then there is a chance that the incident can result in a false positive.

The following are some guidelines that play a critical role in deciding when to declare an incident, starting with *defining thresholds*.

Defining thresholds

Every organization should define a clear set of thresholds in order to declare an event as an incident. These thresholds can be defined by setting up alerts. A specific event can be triggered by a single alert or a combination of alerts. An incident can also be triggered based on the total number of active alerts or the duration of an active alert.

The next guideline to evaluate is *assessing impact*.

Assessing impact

Once guidelines are set, thresholds are defined and, as the alerts are fired, it is important to assess the impact of an event. There are multiple aspects involved in assessing the impact, such as the nature of the event, the impact to the end user, the eventual impact to the business, any dependent stakeholders at risk, or any financial loss to the business.

The following is a reference template on how an incident can be classified based on possible impact:

- **Negligible**: Little or no impact on product but might require follow-up action items
- **Minor**: Impact to internal users but external users might not notice

- **Major**: Impact to external users with noticeable revenue loss but services are still functioning
- **Huge/disaster/detrimental**: Outage severely impacting users and business with a significant revenue loss

At a high level, the following questions can help to determine whether an event is an incident. If the answer is *yes* to any of the following, an incident can be declared:

- Is the outage visible to users?
- Do SREs require the expertise of another team in order to evaluate the incident?
- Does the issue exist after an hour?

Once an event is identified as an incident and its severity level has been identified, SREs or operators should formally declare the incident by issuing a statement that includes the severity level, a list of services impacted, and a possible estimated time of recovery. The summarizes the topic related to *declaring an incident* along with a few essential guidelines. The next critical element to discuss related to effective incident management is *separation of responsibilities*.

Separation of responsibilities

Once an incident has been declared, an incident response team should be formed whose main task is to mitigate or to resolve the incident. The members of the incident response team should have well-defined responsibilities. SRE prescribes a specific set of roles that should be designated within the team. The prescribed roles are as follows:

- **Incident Commander (IC)**
- **Communications Lead (CL)**
- **Operations Lead (OL)**
- **Primary/Secondary Responder**
- **Planning Lead (PL)**

Each of the preceding roles should have a clear charter and autonomy and the same is elaborated in the upcoming sub-sections, starting with the IC.

Incident Commander (IC)

The responsibilities of the IC are as follows:

- The IC leads the chain of command and designates specific roles to specific members of the team. Every member of the incident response team reports to the IC.

- The IC is and should be aware of significant events during an incident response, actively co-ordinate the response during an incident, decide priorities, and delegate activities.

- The IC's core goal is to ensure the problem is mitigated and ultimately fixed. However, the IC does not personally or individually fix the problem.

- The IC initiates the postmortem report after an incident is mitigated or resolved.

- Based on the size of the team, the IC can also assume the role of the CL.

The next topic details the responsibilities of the CL.

Communications Lead (CL)

The responsibilities of the CL are as follows:

- The CL is the public face of the incident response and leads the communication with the outside world, provides timely updates, and also takes questions related to the incident.

- The CL acts as a shield and avoids direct communication between the customer/client and other members of the incident response team.

- The CL maintains the live incident state document, which is used later for postmortem.

The next topic details the responsibilities of the OL.

Operations Lead (OL)

The responsibilities of an OL are as follows:

- The OL develops and executes the incident response plan, thereby being responsible for the technical and tactical work.

- The OL is responsible for the operations team, which comprises Primary and (optionally) Secondary Responders.

- The OL is always in contact with the IC and the CL. If required, the OL requests additional resources either for the operations teams or subject matter experts based on the specific nature of the incident.

The next topic details the responsibilities of the Primary/Secondary Responder.

Primary/Secondary Responder

The responsibilities of a Primary/Secondary Responder are as follows:

- Primary and (optionally) Secondary Responders are members of the operations team and report to the OL. Their main goal is to execute the OL's technical response plan for the incident.

- The Secondary Responder is there to help the Primary Responder or can be designated a particular task or function based on the OL.

The next topic details the responsibilities of the PL.

Planning Lead (PL)

The responsibilities of the PL are as follows:

- The PL works with the operations team and tracks system changes and arranges hand-offs as necessary.

- The PL arranges hand-offs between teams and also reports bugs to keep track of long-term changes.

Figure 4.2 illustrates how the above mentioned roles can be structured:

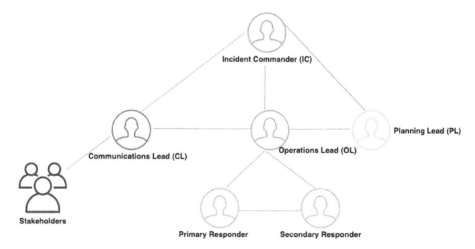

Figure 4.2 - Illustrating Incident Management Roles

Having well-defined roles allows SRE engineers to have a specific area of focus and also prevents outside ad hoc intervention. It is critical to establish a communication post, which could be a physical location such as a meeting room or a communication channel such as Slack. In addition, implementing a clear real-time hand-off either at the end of a shift or the day is highly recommended to ensure the incident is handed off explicitly to another incident commander.

Summarizing the principles of incident response

The core principles include the need to maintain a clear chain of command, designate well-defined roles, maintain a live incident state document, ensure timely communication to the impacted parties about the incident state, perform a live hand-off to avoid operational overload, prepare a postmortem report to determine the root cause, update playbooks as needed, and plan to perform **Disaster Recovery** (**DR**) exercises as often as possible.

The next critical element to discuss related to effective incident management is *recommended best practices*.

Recommended best practices

The following is a summary of best practices that are recommended during the incident management process:

- Develop and document procedures.

- Prioritize damage and restore service.

- Trust team members in specified roles.

- If overwhelmed, ask for help.

- Consider response alternatives.

- Practice procedure routinely.

- Rotate roles among team members.

In the next topic, we will see how we can restore the service after an incident occurs and prevent such incidents from happening in the future.

Restoring service and avoiding recurrence

Once an incident has been declared, the focus of the incident response team should be to troubleshoot the incident. This could be initiated by thinking through the various steps involved in the functioning of the service and stepping through the inner details. The incident can be caused either by internal or external factors. Both kinds of factors should be examined.

While examining internal factors, the focus should be on analyzing any recent code or configuration changes in the area of impact. In most scenarios, a code or configuration change can be reverted, which would eventually restore the service. But there will be situations where multiple code or configuration changes cannot be reverted from a deployment standpoint and the only way to move forward is to provide a code-level fix in order to revert the changes. In contrast to internal factors, examining or dealing with external factors is more complicated, as in most cases there is little control over them.

Irrespective of the nature of the issue, the primary goal is to find ways to resolve or mitigate the issue. Once the issue is mitigated or resolved, a postmortem should be conducted, with the intention to identify the root cause. The process should happen in a blameless fashion with the sole intention to find ways that could prevent the incident from re-occurrence in the future.

The postmortem process should result in a postmortem report that essentially outlines the events of the incident and consists of details with respect to the nature of the impact, the possible root cause, the triggering event, the metrics that help in identifying the event, and a list of action items. In addition, the postmortem report should provide clarity that can help in future mitigation and can be used as use case scenarios (after the incident). This further helps to promote learning among teams.

This completes an in-depth section on incident management. We started the section by trying to outline the life cycle of an incident (*Figure 4.1*) and then elaborated on key constructs of effective incident management, charted the life cycle of an incident, detailed the possible roles and their respective responsibilities and recommended a set of best practices.

Essentially, it is important to differentiate when to call out an incident, identify the impacted parties (internal or external users), assess the level of impact, have predefined roles if an incident is called, and attempt to stick to the charter. This avoids ambiguity in communication and collaboration. This also allows you to focus on the key goal, which is to mitigate the issue and restore the service at the earliest. It is also critical to investigate, identify, and address the root cause to avoid re-occurrence.

The next section focuses on the process of being on call (another SRE cultural practice) and the critical factors that need to be considered.

Being on call

On call duty refers to specific operational activities performed to support the reliable running of a service or application both during working and non-working hours. On call duty is one of the critical responsibilities for an SRE and is also important from a service standpoint to keep the service available and reliable. SRE teams (as previously defined) are different from regular operational teams as the goal is to emphasize the use of engineering techniques to solve operational problems and to prevent their occurrence at scale. It is typically common to engage the product/application SRE team during on call. In the case of specialized services, embedded SRE teams are engaged for on call duty.

When an incident occurs, the response time from initiating the incident management process to resolving or mitigating the issue is key to meeting the desired SLO, which in turn will meet the promised SLA. There are multiple factors that need to be considered while implementing the on call process. The following are three such considerations:

- Paging versus non-paging events
- Primary versus secondary on call rotation
- Single-site versus multi-site production teams

The next topic discusses *paging versus non-paging events*, one of the key factors in implementing the *on call* process.

Paging versus non-paging events

Paging events refers to higher-priority alerts that require immediate attention or remediation, especially in the case of a user-facing application. Examples of paging events could be scenarios where the service health check fails, or a database tied to the service is unable to accept any more connections.

Non-paging events refer to lower-priority alerts that might not point to a service failure but point to issues that need to be addressed before they snowball into a bigger incident. Example of non-paging events include a sudden spike in traffic due to a new feature release in a specific region or a lower-priority ticket raised by a self-healing process when disk usage went up to 80% but then was mitigated automatically by an automated process that increased the disk space by an extra 10%. This allows enough time for the on call engineer to investigate the root cause on the disk usage spike in the first place.

Paging events are always the top priority for the SRE team. The first step in such a scenario is to validate whether the event is an incident and take appropriate steps to initiate the incident management process. In addition, SRE teams also vet non-production paging events and handle those events during their business hours based on operational load. It is important to differentiate between paging and non-paging events. This differentiation needs to be factored while configuring alerting rules, or else it will lead to alert fatigue. Alert fatigue creates a tendency among team members to ignore important alerts.

The next topic discusses *primary versus secondary on call rotation*, one of the key factors in implementing the *on call* process.

Primary versus secondary on call rotation

There could be multiple SRE teams that are on call at a given time. Given that the key responsibility of the on call SRE team is to reliably run the service by tackling both paging and non-paging events that include handling alerts, tickets, and operational duties, it is often common to divide the SRE teams into primary and secondary teams.

This division into primary and secondary teams essentially helps to distribute duties and organize priorities. The distribution of duties between the teams can differ from the way an organization implements SRE. One implementation is to deploy the secondary team as a fall-through for the pages that the primary on call teams cannot get to, potentially because the primary on call team is actively engaged. Essentially, the secondary team is used as a contingency to the primary team in situations where there are more paging events than the primary team can handle at that moment. Another implementation is to assign the primary on call team to always work on paging events while the secondary on call team can work on non-paging production activities.

The next topic discusses *single-site versus multi-site production teams*, one of the key factors in implementing the *on call* process.

Single-site versus multi-site production teams

A single-site production team refers to one or more SRE teams supporting operations from a single location. Though it's easy for the teams to communicate and exchange information, the challenge is that specific teams have to be engaged during night shifts, which could be detrimental in the long term from a health standpoint.

A multi-site production team refers to one or more SRE teams supporting operations from multiple locations. The typical approach is to ensure the locations are in geographically different regions with a *follow-the sun* rotation model that allows the teams to completely avoid night shifts.

The *follow-the-sun* rotational model is a common term used to represent a service and support methodology where a user-facing application is supported by multiple teams that are spread across the world to provide 24/7 support rather than forcing a single support team at a specific location to work overtime till the issue is resolved. In the case of this model, if an outstanding issue is not resolved beyond the shift time of a specific team, then the issue will be transitioned to the team that will handle the next shift, along with the detailed analysis and steps taken by the prior team. The downsides of a multi-site production team include communication challenges and co-ordination overhead.

The next topic discusses the recommended practices while being on call.

Recommended practices while being on call

The following are some SRE recommended practices while being on call:

- Dealing while being on call is all about the approach. While the thinking needs to be rational and focused, there is a risk that the actions can be intuitive and heuristic. Such a risk should be avoided.

- Intuitive actions can often be wrong and are less supported by data. Intuition can lead the on call SREs (SRE engineers) to pursue a line of reasoning that is incorrect from the start and could potentially waste time.

- A heuristic approach creates a tendency where on call SREs can take an approach based on assumptions and previous occurrences, but the approach might not be optimal or rational in nature.

- The most important facets while being on call are to have a clear escalation path, a well-defined incident management procedure, and a blameless postmortem culture, and to strike a balance between *operational overload* and *operational underload*.

- Develop a culture of effective troubleshooting by trying to fundamentally understand how a system is built, designed, and supposed to work. Such expertise can be gained by investigating when a system doesn't work. In addition, there should be a constant focus on asking questions, with constructs such as *what*, *where*, *why*, and *how* that could potentially lead to the next set of connected questions or answers and eventually help to troubleshoot an issue.

- Avoid common pitfalls such as focusing on symptoms that aren't relevant, a lack of understanding about the impact of a misconfiguration, and attempting to map a problem to other problems in the past by drawing correlations or using assumptions that are inaccurate.

> **Operational overload versus operational underload**
>
> Operational overload is a state that is caused by a misconfigured monitoring system or incorrect choice of alerting rules, leading to fatigue and stress. This typically leads to an increase in ticket count, pages, and ongoing operational support. Handling high-priority tickets and pages leads to tense situations, causing stress, and could also restrict a Site Reliability Engineer's ability to continue working on their engineering engagements. Such situations can be eased by temporarily loaning an experienced SRE to an overloaded team as dedicated help, providing breathing space to teams to allow them to address issues or allow the team member to focus on engineering projects.
>
> Operational underload is a state where a Site Reliability Engineer is not involved with a production system for a long period of time. This could create significant knowledge gaps and a lack of confidence while suddenly dealing with a production issue. This can be avoided if SRE teams are sized such that every engineer is on call at least once or twice in a quarter.

This completes the section on the process related to being on call, factors that could have an impact while implementing the process, and recommended best practices. The next section focuses on the cultural SRE practices of psychological safety and factors to achieve the same.

Psychological safety

One of the key pillars of SRE is to accept failure as normal. This implies that failures are imminent, but the key is to learn from the failures and ensure that the same mistake is not repeated the next time. As a result, SRE promotes open communication within teams and between members of the team and its leaders to ensure that a failure is evaluated objectively from a process standpoint and not from an individual standpoint. The core idea is to provide a sense of psychological safety, which is extremely important to implement the practice of **Blameless Postmortems**.

SRE defines psychological safety as the belief that a manager or an individual contributor will not be singled out, humiliated, ridiculed or punished for the following:

- Committing a mistake that could result in a potential incident or a problem

- Bringing up a concern related to a decision with respect to design, implementation, or process that could later have adverse impacts

- Asking questions for further clarifications that could help the individual or multiple members of the team to effectively implement or collaborate

- Coming up with newer ideas that could foster innovation for further improvement of a process or service

> **Disadvantages of low psychological safety**
>
> A workplace or a team with low psychological safety will eventually suppress learning and innovation. Team members will be apprehensive about clarifying or speaking out or initiating a conversation. It creates a sense of self-doubt and impacts the exchanging of ideas that could lead to innovation.

Psychological safety is also important in incident management. In the following topic, we will discuss some factors to ensure psychological safety.

Factors to overcome in order to foster psychological safety

A thought process of blamelessness promotes psychological safety, and the following two factors should be overcome to avoid blame in the first place:

- **Hindsight bias**: Refers to a know-it-all attitude that something will eventually fail even though it was not obvious at that point of time. As a result, there is a tendency to blame the person leading the effort that the required actions were not taken to foresee the issue and avoid the failure.

- **Discomfort discharge**: Refers to a tendency where people blame to discharge discomfort. This leads to a tendency where team members tend to conceal information, facts, or not communicate because of a fear of punishment or consequences that could negatively impact the individual or the team.

To conclude, psychological safety is critical to allow team members to have the freedom to take or make decisions based on the best possible information available at that point in time. In addition, and more often than not, innovation always includes a need to take risks, which also means there is a chance of failure. In either case, if something goes wrong, the focus needs to be on the process leading to the failure but not on the people involved during the incident. This allows for an objective analysis and provides freedom for the team to express their thoughts without hesitation, leading to open communication and continuous improvement.

Head, heart, and feet model

The head, heart, and feet model is a way to promote psychological safety from the fear of adopting a change. The head represents the rational, where the emphasis must be on why the change is happening and should include the strategic mission and vision. The heart represents the emotional, where the emphasis should be on how change can bring in a positive impact. The feet represent the behavioral, where the emphasis should be on the knowledge, skills, and resources that should be provided to implement change successfully.

The next section deep-dives into another set of SRE cultural practices that promotes sharing vision and knowledge and fostering collaboration.

Sharing vision and knowledge and fostering collaboration

One of the key pillars of SRE is to reduce organizational silos, and this can be achieved by creating a unified vision, sharing knowledge, and fostering collaboration.

Unified vision

Organizations have a vision statement that serves as a guide for the work they do or represent. A team's vision should be in line with the organization's vision and typically this vision will have the following constructs:

- **Core values**: Helps teams to build trust and psychological safety, creates a willingness to take risks and be open to learn.

- **Purpose**: Refers to the specific intent for the existence of the team.

- **Mission**: Points to a clear and compelling goal that the team strives to achieve.

- **Strategy**: Refers to the path to realize or achieve team's mission; this includes the process to identify relevant resources, capabilities, threats, and opportunities.

- **Goals**: Refers to a defined set of team's objectives. SRE recommends the use of OKRs to set ambitious goals with a drive to accomplish more than possible; **OKRs** refer to **objective and key results**, a collaborative goal-setting framework that is used by individuals and teams to aim for ambitious goals and measure results during the process.

OKRs can enable teams to focus on big bets and accomplish more than the team thought was possible, even if they don't fully attain their intended goal. This is accomplished by clearly defining an objective and also classifying key results quantitatively to ensure that the objective is achieved. The objective of setting OKRs is to set a minimum of those defined key results (if not all). OKRs can encourage people to try new things, prioritize work, and learn from both successes and failures. While the team may not reach every OKR, it gives them something to strive for together, driving toward a unified vision.

Communication and collaboration

Communication and collaboration between teams is critical to implementing SRE. This could include the communication between multiple SRE teams within the organization or communication between SRE teams and their respective product/development team. This leads to identifying common approaches to solve problems that might have common elements, remove ambiguity, and provides the possibility to solve more complex challenges. The following are some options proposed by the SRE team.

Service-oriented meetings

Let's understand what these meetings are for:

- These are mandatory meetings that are meant to review the state of service and increase awareness among all possible stakeholders.

- The recommended duration is about 30-60 minutes and should be driven by a designated lead with a clearly defined agenda.

Next, we will discuss team composition.

Effective team composition

Let's see how effective team composition is achieved:

- SRE recommends that every SRE team should have an effective team composition – specifically on certain roles such as Technical Lead, Manager, and Project Manager.

- A Technical Lead sets the technical direction of the team. A manager runs the team's performance management and is the first point of contact for the team. A Project Manager comments on design documentation.

Another approach for communication and collaboration is knowledge sharing.

Knowledge sharing

Here's what this approach entails:

- Cross-training, an employee-to-employee network, and job shadowing are the most recommended options.

- Cross-training allows you to increase competencies of a team member by training in other competencies, thus encouraging employees to constantly learn and grow.

- An employee-to-employee network encourages employees to share their knowledge and learning by driving information sessions internally.

- Job shadowing refers to on-the-job training by observing and potentially helping personnel in their area of expertise.

This completes the final section on a key SRE cultural practice that focuses on sharing a unified vision and knowledge, and fostering communication and collaboration. This also brings us to the end of this chapter.

Summary

In this chapter, we discussed the key elements that are required to build an SRE team. In addition, we discussed key cultural practices that help to implement the technical practices. This concludes the first section of the book focussed on SRE (*Chapters 1-4*). The next set of chapters (*Chapters 5-10*) will focus on the core Google Cloud services required to implement DevOps, starting with *Chapter 5, Managing Source Code using Cloud Source Repositories.*

Points to remember

The following are some important points to remember:

- IC: In charge of the incident response, designates responsibilities and optionally takes on roles that were not designated, such as CL.

- OL: Deals with resolving or handling the incident – technical and tactical; executes the action plan by working with a Primary and Secondary Responder.

- CL: The public face of incident response; responsible for communicating to all the stakeholders.

- PL: Tracks system changes, identifies long-term changes by filing bugs, and arranges hand-offs.

- Command post: Refers to a meeting room or Slack channel where communication happens.

- Live incident state document: Maintained by the CL about the incident and updates, and is later used for the postmortem.

- An incident or outage should be called if specific expertise is needed, if the outage is visible, or if the issue is not resolved after an hour or so of effort.

- Factors to overcome in order to foster psychological safety: hindsight bias and discomfort discharge.

- A team's vision is everything about what drives its work and includes the core values, purpose, mission, strategy, and goals.

- A team's mission is a clear and compelling goal that it wants to achieve.

Further reading

For more information on GCP's approach toward DevOps, read the following articles:

- **SRE**: `https://landing.google.com/sre/`

- **SRE fundamentals**: `https://cloud.google.com/blog/products/gcp/sre-fundamentals-slis-slas-and-slos`

- **SRE YouTube playlist**: `https://www.youtube.com/watch?v=uTEL8Ff1Zvk&list=PLIivdWyY5sqJrKl7D2u-gmis8h9K66qoj`

Practice test

Answer the following questions:

1. As per the SRE engagement model, during which phase do SREs define SLOs?

   a) Architecture and design

   b) Active development

   c) Limited availability

   d) General availability

2. Who out of the following is responsible for initiating a postmortem report after an incident?

   a) **Incident Commander (IC)**

   b) **Communications Lead (CL)**

   c) **Operations Lead (OL)**

   d) **Planning Lead (PL)**

3. As per the SRE engagement model, during which phase is SREs' engagement deepest or highest?

   a) Architecture and design

   b) Active development

   c) Limited availability

   d) General availability

4. As per the SRE engagement model, during which phase do SREs start preparing the service for production?

   a) Architecture and design

   b) Active development

   c) Limited availability

   d) General availability

5. Who out of the following is in charge of responding to an outage or an incident?

   a) **Incident Commander (IC)**

   b) **Communications Lead (CL)**

   c) **Operations Lead (OL)**

   d) **Planning Lead (PL)**

6. Who out of the following is in charge of executing the technical response for an outage or an incident?

   a) **Incident Commander (IC)**

   b) **Communications Lead (CL)**

   c) **Operations Lead (OL)**

   d) **Primary Responder (PR)**

7. Select the incident severity classification that has the following characteristics: the outage is visible to the user with a noticeable revenue loss but no lasting damage.

    a) Negligible

    b) Minor

    c) Major

    d) Detrimental

8. Who out of the following is in charge of managing the immediate, detailed technical and tactical work of the incident response?

    a) **Incident Commander (IC)**

    b) **Communications Lead (CL)**

    c) **Operations Lead (OL)**

    d) **Planning Lead (PL)**

9. Who out of the following is in charge of coordinating the efforts of the response team to address an active outage or incident?

    a) **Incident Commander (IC)**

    b) **Communications Lead (CL)**

    c) **Operations Lead (OL)**

    d) **Planning Lead (PL)**

10. Select the incident severity classification that has the following characteristics: little or no impact on production but might require low-priority follow-up actionable items.

    a) Negligible

    b) Minor

    c) Major

    d) Detrimental

Answers

1. (c) Limited availability
2. (a) Incident Commander
3. (a) Architecture and design
4. (b) Active development
5. (a) Incident Commander
6. (d) Primary Responder
7. (c) Major; because of revenue loss but no lasting damage
8. (c) Operations Lead
9. (a) Incident Commander
10. (a) Negligible

Section 2: Google Cloud Services to Implement DevOps via CI/CD

The core focus of this section is to deep dive into Google Cloud services that are critical to manage source code using Cloud Source Repositories, build code and create build artifacts, push artifacts to a registry, deploy the artifacts as containerized applications, orchestrate these applications in a cluster through managed compute services, secure the clusters, and finally, implement observability on the deployed applications through a suite of services focused on operations.

The section introduces key features around each of the core services along with hands-on labs as needed. All the labs throughout the section are connected, to provide a holistic understanding. The section concludes with SLO monitoring, a feature in Cloud Operations that allows tracking SRE technical practices for an application deployed to **Google Kubernetes Engine (GKE)**.

This section comprises the following chapters:

- *Chapter 5, Managing Source Code Using Cloud Source Repositories*
- *Chapter 6, Building Code Using Cloud Build, and Pushing to Container Registry*
- *Chapter 7, Understanding Kubernetes Essentials to Deploy Containerized Applications*
- *Chapter 8, Understanding GKE Essentials to Deploy Containerized Applications*
- *Chapter 9, Securing the Cluster Using GKE Security Constructs*
- *Chapter 10, Exploring GCP Cloud Operations*

5
Managing Source Code Using Cloud Source Repositories

The first section of this book (consisting of four chapters) focused on exploring the concepts of **Site Reliability Engineering** (**SRE**) in depth. This included SRE technical practices, understanding monitoring and alerting to target reliability, and insights into building SRE teams by applying SRE cultural practices.

This second section of the book explores GCP's constructs in depth to implement a CI/CD pipeline with a focus on the following core areas:

- Managing source code using Cloud Source Repositories
- Building and creating container images using Cloud Build
- Pushing container images and artifacts using Container Registry
- Orchestrating containers and deploying workloads using Google Kubernetes Engine

Source code management is the first step in a **Continuous Integration** (**CI**) flow. Code is stored in a source code repository; a common repository that stores code and allows developers to make code changes (if required in isolation) and merge changes from multiple contributors into a single common stream. The most common examples of a source code repository include GitHub and Bitbucket. **Cloud Source Repositories** (**CSR**) is a service from Google Cloud that provides the functionality of source code management through private Git repositories and easily integrates to several Google Cloud services such as Cloud Build, Cloud Monitoring, and Cloud Logging.

In this chapter, we're going to cover the following main topics:

- **Key features**: Managed private Git repository, one-way sync with external repositories, universal code search, and native integration with other **Google Cloud Platform** (**GCP**) services.

- **First steps**: Create a first repository via the console or the **Command-Line Interface** (**CLI**) and add files to the repository.

- **One-way sync from GitHub/Bitbucket to CSR:** Option to add a repository by connecting to an external repository and perform near real-time one-way sync.

- **Common operations**: Browse repositories, browse files, perform universal code search, detect security keys, and assign the right access controls.

- **Hands-on lab**: Step-by-step instructions to integrate CSR with Cloud Functions.

Technical requirements

There are four main technical requirements:

- A valid GCP account to get hands-on with GCP services: `https://cloud.google.com/free`.

- Install Google Cloud SDK: `https://cloud.google.com/sdk/docs/quickstart`.

- Install Git: `https://git-scm.com/book/en/v2/Getting-Started-Installing-Git`.

- Alternatively, it is possible to skip the previous two and instead install Cloud Shell, which includes Google Cloud SDK as well as Git: `https://cloud.google.com/shell`.

Introducing the key features

CSR is a service from Google Cloud to manage source code. CSR provides Git version control and supports the collaborative development of any application or service. Key features include the following:

- **Fully managed private Git repository**: This feature implies that there is no need to manage the infrastructure required to host the source code repository. Developers can instead focus on building, testing, deploying, and debugging code.

- **Provides one-way sync with GitHub and Bitbucket**: In situations where developers use either GitHub or Bitbucket as their primary cloud source repository, enabling integration with other GCP services such as App Engine, Cloud Functions, Cloud Monitoring, and Cloud Logging is not as straight forward in comparison to using GCP's CSR. For example, it easy to deploy code to a serverless service in GCP such as Cloud Functions from CSR directly then GitHub or Bitbucket instead. In addition, the one-way sync feature performs a one-way mirror, essentially making a near real-time copy of a repository from GitHub or Bitbucket into GCP's CSR. This facilitates the ease of native integration with GCP services.

- **Includes universal code search**: This feature allows you to perform a code search within the code source repository or across repositories. Search can also be scoped to a specific project or repository or even a specific directory.

- **Integration with GCP services**: This feature allows native integration with multiple GCP services such as Cloud Build, Cloud Operations, Cloud Functions, and Cloud Run. For example, the logs related to operations against CSR are automatically sent to Cloud Logging. The user, however, requires relevant **Identity Access Management (IAM)** roles to access Cloud Logging in order to view logs related to CSR.

> **Identity Access Management (IAM)**
>
> IAM is a framework of roles and policies to ensure users and applications have the required access to resources specifically recommended within the principles of least privilege.

Each of the previously mentioned features will be discussed in detail later in this chapter. The upcoming section details the step-by-step process involved to create and access a repository in CSR.

First steps – creating and accessing a repository in CSR

One of the first steps to perform while working with CSR is to actually create a repository and add files to it. Given that CSR is a managed repository, the user need not manage the space required to host the repository or the computational power required to maintain or run operations against the repository.

In this section, we will see how we can create a repository in CSR from Google Cloud Console and the command line. In addition, we'll learn how to add files to a branch in an empty repository and then merge to the master. Let's get started.

Creating a repository via Google Cloud Console

The following is a step-by-step process to create our first repository in CSR through Google Cloud Console:

1. Enable the CSR API (*Figure 5.1*) by navigating to the **Library** sub-section under the **APIs & Services** section:

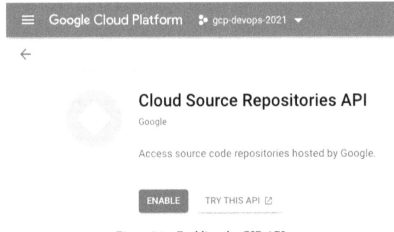

Figure 5.1 – Enabling the CSR API

2. Navigate to **Source Repositories** within GCP and select the **Get Started** option. The system will display a prompt (*Figure 5.2*) and provide an option to create a repository. However, if a repository already exists in CSR, skip to *step 3* and use the **Add a Repository** option instead:

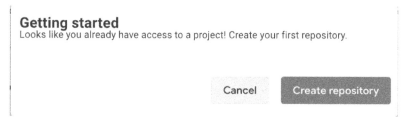

Getting started
Looks like you already have access to a project! Create your first repository.

Cancel   Create repository

Figure 5.2 – Option to create your first repository

3. The system will prompt to add a repository by providing two options (*Figure 5.3*) – to either create a new repository or connect to an external repository. In this case, select the option to create a new repository:

Add a repository

Select one of the following options to continue:

⦿ Create new repository
Choose this option to create an empty repository.

◯ Connect external repository
Choose this option to mirror a repository from a hosted service, such as GitHub or Bitbucket.

Cancel   Continue

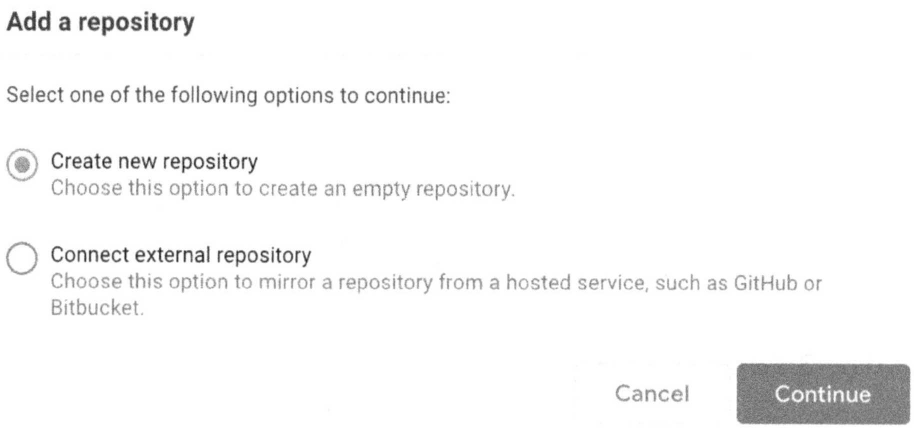

Figure 5.3 – Option to create a new repository

4. Create a repository by entering a repository name. Additionally, select a project under which the repository should be created (*Figure 5.4*):

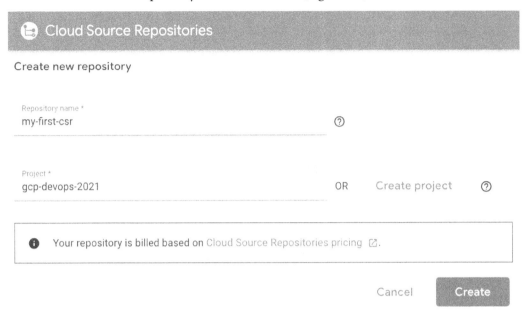

Figure 5.4 – Creating a new repository

This will create a new repository from the console. However, there will be situations where it is required to create a repository through scripts and that means through the command line. This is specifically recommended when automation is at the forefront and the goal is to eliminate toil. The upcoming topic details how to create a repository via the CLI.

Creating a repository via the CLI

To create a cloud source repository via the command line or CLI, execute the following commands. It is required to either install Google Cloud SDK or use Google Cloud Shell:

```
# Enable the Cloud Source Repository API
gcloud services enable sourcerepo.googleapis.com

# Create a repository
gcloud source repos create my-first-csr --project $GOOGLE_
CLOUD_PROJECT ops-2021
```

The preceding steps will create a new repository from the CLI. Essentially, the new repository created in CSR from either the console or the CLI will be an empty repository. The next topic will detail how to add files to a repository in CSR.

Adding files to a repository in CSR

Once a repository is created, developers can create a branch and make their changes inside that branch. These changes can then be merged into the master once confirmed. This is a multi-step process (as detailed in the following procedure) that could be executed from the user's terminal window with Google Cloud SDK installed or via Google Cloud Shell from the user's choice of browser:

1. Clone the repository to a local Git repository:

    ```
    gcloud source repos clone my-first-csr --project=$GOOGLE_
    CLOUD_PROJECT
    ```

2. Switch to the new local Git repository:

    ```
    cd my-first-csr
    ```

3. Create a new branch:

    ```
    git checkout -b my-first-csr-branch
    ```

4. Add a file to the new branch:

    ```
    touch hello.txt
    git add hello.txt
    ```

5. Commit changes of the new file to the branch:

    ```
    git commit -m "My first commit!!"
    ```

6. Push changes to the branch:

    ```
    git push --set-upstream origin my-first-csr-branch
    ```

7. Create a master branch (as this is the first check into `master`):

    ```
    git checkout -b master
    ```

8. Merge the branch to `master`:

    ```
    git push --set-upstream origin master
    ```

This completes this section and you can now create an empty repository and sub-sequently add files to a working branch and then check into the master. However, there will be situations where the user can work with either an existing repository in GCP's CSR or external source repositories in GitHub/Bitbucket. Either way, the process to clone an existing repository is the same. In addition, CSR allows one-way sync from external repositories such as GitHub/Bitbucket. All these details will be covered in the next section.

One-way sync from GitHub/Bitbucket to CSR

CSR provides an option to add a repository by connecting to an external repository and perform near real-time one-way sync. Currently, GitHub and Bitbucket are the only supported external source repositories.

The following is the step-by-step process to create a repository in CSR by connecting to an external GitHub repository (similar steps will apply to a Bitbucket-based repository):

1. Navigate to **Source Repositories** in Google Cloud Console and select the **Add Repository** option.

2. Select the option to connect to an external repository (*Figure 5.5*):

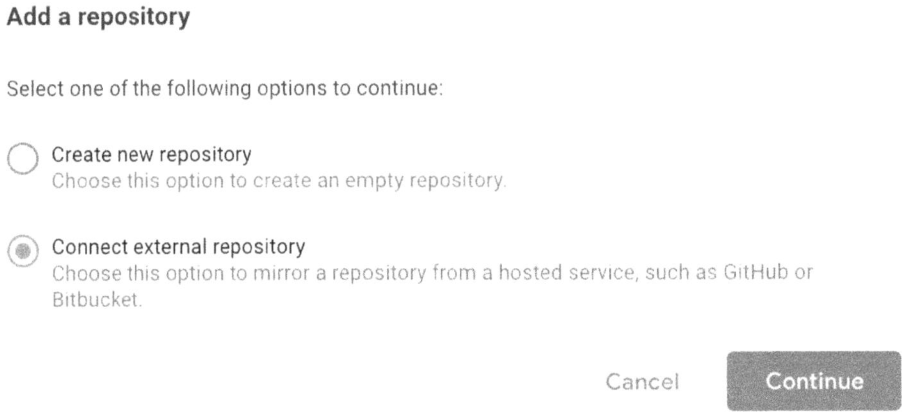

Add a repository

Select one of the following options to continue:

○ Create new repository
  Choose this option to create an empty repository.

◉ Connect external repository
  Choose this option to mirror a repository from a hosted service, such as GitHub or Bitbucket.

Cancel   Continue

Figure 5.5 – Option to connect to an external repository

3. Select an appropriate project and an external Git provider (**GitHub** in this case) and authorize the selected GCP project to store third-party authentication credentials in order to enable connected repository services (*Figure 5.6*):

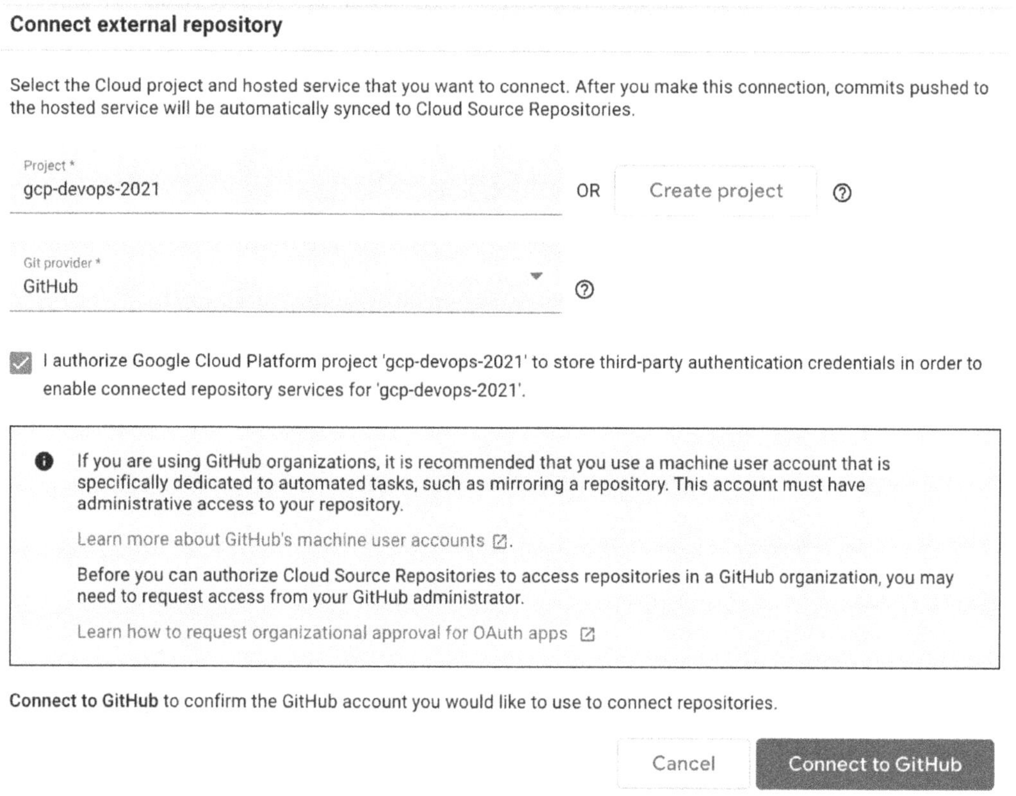

Connect external repository

Select the Cloud project and hosted service that you want to connect. After you make this connection, commits pushed to the hosted service will be automatically synced to Cloud Source Repositories.

Project *
gcp-devops-2021   OR   Create project   ⑦

Git provider *
GitHub   ▼   ⑦

☑ I authorize Google Cloud Platform project 'gcp-devops-2021' to store third-party authentication credentials in order to enable connected repository services for 'gcp-devops-2021'.

> ❶ If you are using GitHub organizations, it is recommended that you use a machine user account that is specifically dedicated to automated tasks, such as mirroring a repository. This account must have administrative access to your repository.
>
> Learn more about GitHub's machine user accounts ☐.
>
> Before you can authorize Cloud Source Repositories to access repositories in a GitHub organization, you may need to request access from your GitHub administrator.
>
> Learn how to request organizational approval for OAuth apps ☐

Connect to GitHub to confirm the GitHub account you would like to use to connect repositories.

Cancel   **Connect to GitHub**

Figure 5.6 – Connecting to GitHub

4.   Enter your GitHub credentials and authorize GCP to access the provided GitHub account (*Figure 5.7*):

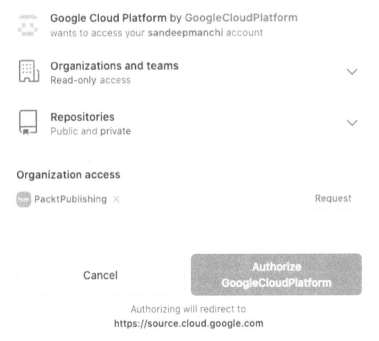

Figure 5.7 – Authorizing GCP to access GitHub

5. Once authorized, select the GitHub repo that needs to be synced with CSR and then select the **Connect selected repository** action (*Figure 5.8*):

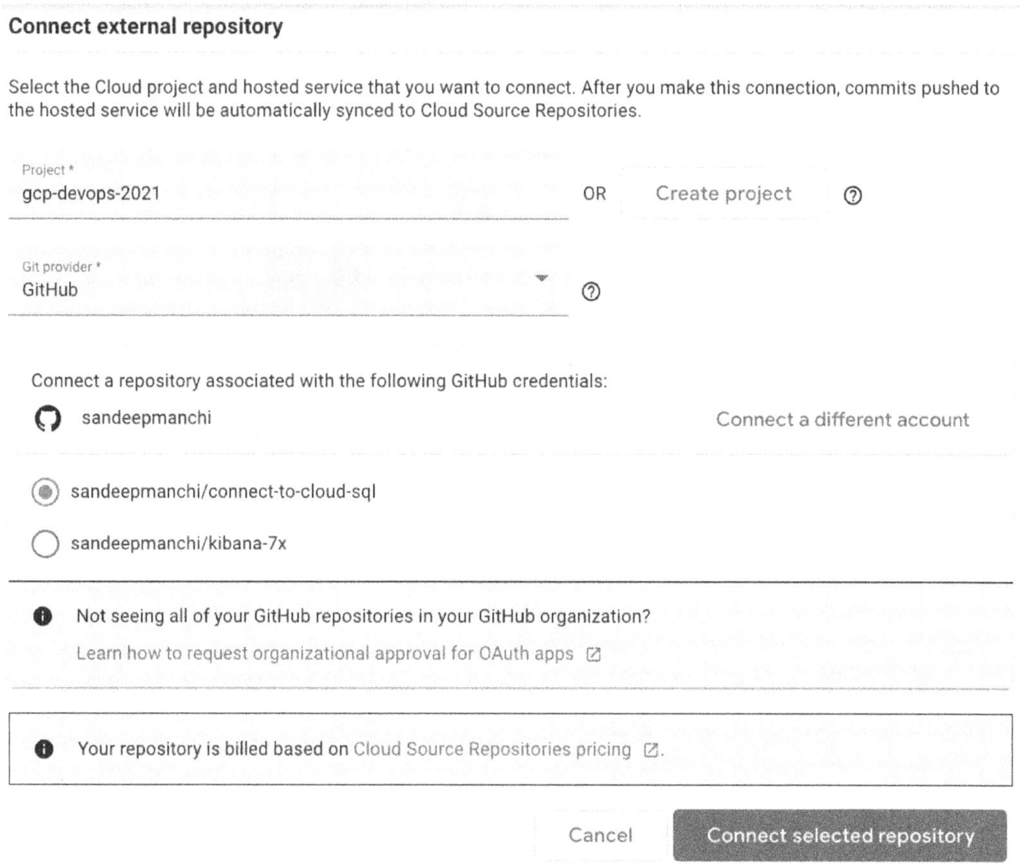

Connect external repository

Select the Cloud project and hosted service that you want to connect. After you make this connection, commits pushed to the hosted service will be automatically synced to Cloud Source Repositories.

Project *
gcp-devops-2021   OR   Create project   ⑦

Git provider *
GitHub   ▼   ⑦

Connect a repository associated with the following GitHub credentials:

○ sandeepmanchi   Connect a different account

◉ sandeepmanchi/connect-to-cloud-sql

○ sandeepmanchi/kibana-7x

ⓘ   Not seeing all of your GitHub repositories in your GitHub organization?
    Learn how to request organizational approval for OAuth apps ☑

ⓘ   Your repository is billed based on Cloud Source Repositories pricing ☑.

Cancel   Connect selected repository

Figure 5.8 – Connecting GCR to the selected GitHub repository

6. The following prompt (*Figure 5.9*) will be displayed once the connection is established between the GitHub repo and GCP's CSR. The first-time sync might take some time, but subsequent syncs are near real-time:

> github_sandeepmanchi_connect-to-cloud-sql
> connected
>
> Repository contents can take some time to appear and show up in search results. Learn more.
>
> OK

Figure 5.9 – Confirmation that one-way sync is established with GitHub

7. The contents of the GitHub repo will eventually sync up with CSR and that also includes the commit history and any other available metadata (*Figure 5.10*):

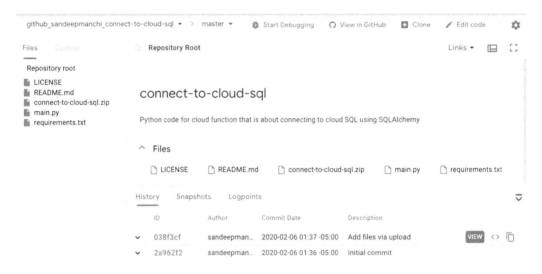

Figure 5.10 – Contents of the newly added GitHub repository

8. If the user adds a new file to the GitHub repo, then a near real-time one-way sync will be performed by CSR. The commit along with the recent changes will reflect in CSR against the relevant project. *Figure 5.11* highlights the new commit history:

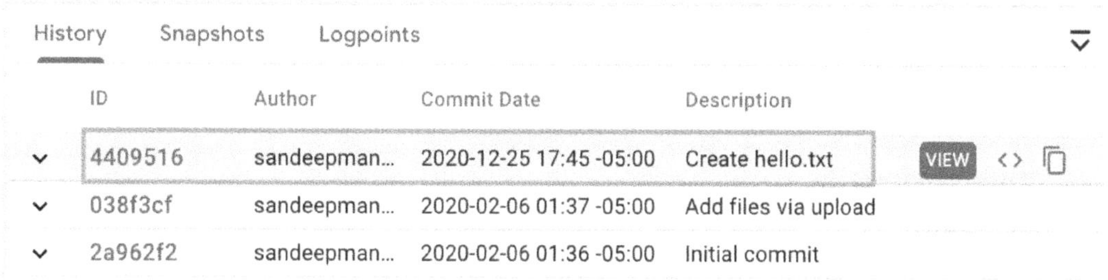

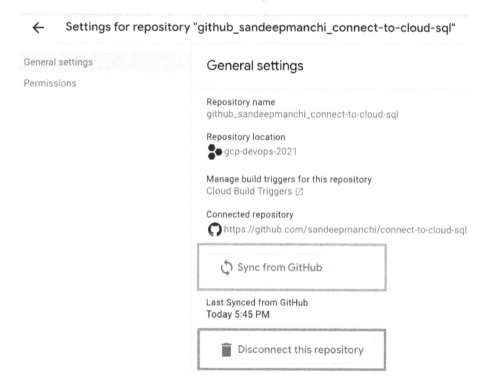

	ID	Author	Commit Date	Description
⌄	4409516	sandeepman...	2020-12-25 17:45 -05:00	Create hello.txt
⌄	038f3cf	sandeepman...	2020-02-06 01:37 -05:00	Add files via upload
⌄	2a962f2	sandeepman...	2020-02-06 01:36 -05:00	Initial commit

Figure 5.11 – Updating the commit history post a near real-time one-way sync

9. If there is a need to perform forced sync from an external repository to CSR or disconnect the external repository from CSR, navigate to the repository settings in GCP to find the appropriate options (*Figure 5.12*):

Figure 5.12 – Repository settings to force a sync or disconnect from GitHub

This completes the detailed step-by-step process of establishing one-way sync with external repositories such as GitHub/Bitbucket. The next section dives into some common operations that a user can perform in CSR, such as browsing repositories and files and performing universal code search.

Common operations in CSR

This section details the common operations that could be performed in CSR. The options include the following:

- Browse repositories.

- Browse files.

- Perform a universal code search.

- Detect security keys.

- Assign access controls.

Let's go through them in detail starting with the browsing repositories option.

Browsing repositories

There are two specific views to browse repositories. These views are represented across two tabs:

- **All repositories**

- **My source**

All repositories

CSR shows a consolidated view of all available repositories across projects that the current user has access to. The combination of repository name and project ID forms a unique tuple.

The user can also mark repositories of choice (typically the most important or most constantly used) with a star. All starred repositories will show up under the **My source** tab to provide quick access to specific repositories (*Figure 5.13*):

Figure 5.13 – List of repositories listed under All repositories

Users can perform three specific operations against any repository displayed under the **All repositories** tab (refer to the green square box in *Figure 5.13*, in the following order):

- **Settings**: This option allows the user to edit settings.
- **Clone**: This option provides details required to clone the repository.
- **Permissions**: This option allows you to control access to a repository either at the level of a user or a group or service account.

The user can access a repository by selecting a repository of choice from the list view or they can pick one from the tree view (via the drop-down control for **All repositories**).

My source

The repositories that are starred in the **All repositories** section are listed to provide quick access to a user-selected subset. Additionally, recently viewed repositories (that may or may not be starred) are also listed and can be accessed with a click (*Figure 5.14*):

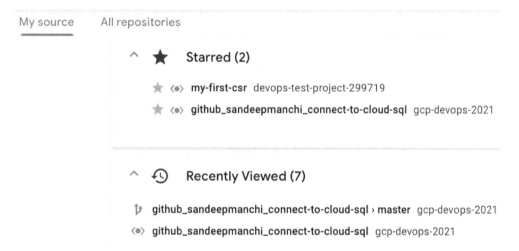

Figure 5.14 – Contents of My source displaying starred and recently viewed repositories

This concludes the details on how a user can browse through the repositories in CSR. The next topic focuses on browsing files within a specific repository.

Browsing files

Once a user selects a repository to browse, the default view switches to the master branch. The user can view the list of files through a tree-like structure (on the left-hand side) and selecting any file will display the contents of the file (on the right-hand side). The user can also edit a file by using the **Edit code in Cloud Shell** option. At this point, the file will be opened in the Cloud Shell Editor (*Figure 5.15*) using the credentials associated with the project. The authentication happens automatically with no additional login:

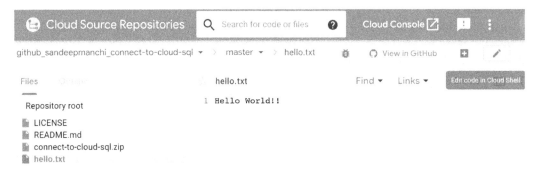

Figure 5.15 – Options to view/edit file contents

The user can also switch to an existing branch by selecting the branch of choice (*Figure 5.16*). In addition, the user can view files by a specific tag or commit:

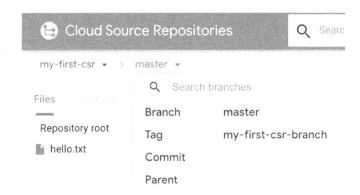

Figure 5.16 – Options to switch branch or browse files by tag or commit

If a user wants to view historical changes for a specific file, then they can view the change information either through the **Blame** panel (on the top right-hand side) or through the **History** sub-section (*Figure 5.17*):

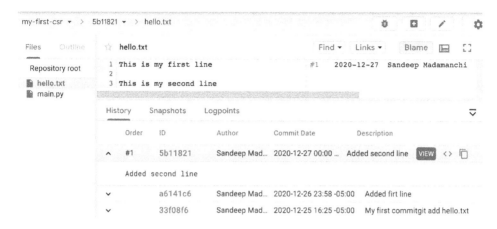

Figure 5.17 – Historical change information for a specific file in CSR

This concludes the details on how a user can browse through files within a specific repository. The next topic focuses on how a user can perform a universal code search within a repository or across repositories in CSR.

Performing a universal code search

CSR provides the ability to search code snippets or files through the search box located on the CSR console. The user can search by typing text (preferably within double quotes) or by using regular expressions.

The scope of the search can be set to one of four possible levels (*Figure 5.18*):

- **Everything**: Search across all repositories that the user has access to.
- **This project**: Search across all repositories in the current project.
- **This repository**: Search across the current repository.
- **This directory**: Search across the current directory:

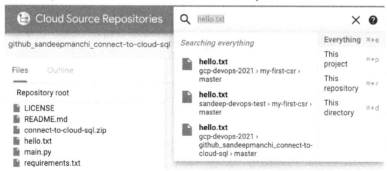

Figure 5.18 – Possible scopes to perform universal code search

The next topic covers the possible ways to perform a code search based on different filter criteria and their respective syntaxes.

Search filters

The following table lists some search filters that can be used to search for code:

Filter Criteria	Syntax	Example

Search file contents	`file:`
	
	`filepath:`
	
	`f:`
		`file:hello.txt`

Search by language	`lang:<language>` `<filename>`	`lang:py main`
Search by function	`function:<function-name> <language>`	`function: execute_db_action py`
Search by excluding terms	`<search-criteria> -<exclude-criteria>`	`function: execute_db_action -lang:py`
Search for literals	`"search-text-in-quotes"`	`"pool_size=5"`

This concludes the details on how a user can perform a universal code search. The next topic focuses on a specific feature in CSR that can specifically enforce a policy to detect security keys when a user attempts to make a code commit.

Detecting security keys

CSR provides options to detect whether security keys are stored in a repository. If this feature is enabled, CSR will enforce this check when a user is trying to push code into the repository either to a branch or master. If the contents of the file include a security key, then the code will not be pushed, and the user will be notified.

Currently, CSR can be set up to check for the following types of security keys:

- Service account credentials in JSON format
- PEM-encoded private keys

The following commands will provide the ability to enable, disable, or override security key detection:

```
# To enable security key detection
gcloud source project-configs update --enable-pushblock
# To disable security key detection
gcloud source project-configs update --disable-pushblock
# To override security key detection at a commit level
git push -o nokeycheck
```

This concludes the details on how security keys can be detected during a code commit. The next topic focuses on the required access controls to perform operations in CSR.

Assigning access controls

Access to repositories can be assigned at either the project level or the repository level. If a user is assigned a specific role at the project level, then that role will be applied to the user for all repositories in that project. However, if a user is assigned a specific role for a specific repository, then it only applies to that repository.

The following table summarizes the critical IAM roles required to access or perform actions on CSR:

Role Name	Role Description
Source Repository Reader	Can list, clone, fetch, and browse repositories
Source Repository Writer	Has Source Repository Reader permissions and can update repositories
Source Repository Admin	Has Source Repository Writer permissions and can create/delete repositories, view/set IAM policies, view/update cloud project configurations, and update repository configurations

The next topic provides information on how cross-account project access can be set up.

Cross-account project access

If a user is part of project A but needs to access a specific repository in project B, then the user should be given the Source Repository Reader/Writer/Admin role, depending on the intended scope of the user from project A against a specific repository in project B. This can be achieved through the **Permissions** section under the repository settings.

This concludes the details on access controls specific to CSR. This also brings us to the end of a major section focused on common operations that users can perform in CSR. The upcoming section is a hands-on lab where a cloud function is deployed using the code hosted in a cloud source repository.

Hands-on lab – integrating with Cloud Functions

The objective of this hands-on lab is to demonstrate the integration between a GCP compute service such as Cloud Functions with CSR. The intent is to illustrate how code can be deployed in Cloud Functions by pulling the code hosted from CSR. The following is a summary of the steps at a high level:

1. Add code to an existing repository through the Cloud Shell Editor.
2. Push code from the Cloud Shell Editor (local repository) into CSR.
3. Create a cloud function and deploy code from the repository in CSR.

Adding code to an existing repository through the Cloud Shell Editor

This sub-section specifically focuses on adding code to an existing repository. Developers typically use their favorite editors to make code changes. The following shows the usage of GCP's Cloud Shell Editor, an online development environment that supports cloud-native development through the Cloud Code plugin along with language support for Go, Java, .NET, Python, and Node.js:

1. Navigate to **Source Repositories** in the GCP console.
2. Navigate to the repository of choice where you want to add code. You can use the `my-first-csr` repository that was previously created.
3. Select the **Edit in Cloud Shell** action. This opens the code in the Cloud Shell Editor and also clones the code from CSR into a local repository under the Cloud Shell console.
4. Add a new code file called `main.py`. Copy the contents of this file from `https://github.com/PacktPublishing/Google-Cloud-Platform-for-DevOps-Engineers/blob/main/cloud-build/main.py`.
5. Save the code file.

Once the code is edited and added, the next step is to push the code into CSR. This will be covered as the next topic.

Pushing code from the Cloud Shell Editor (local repository) into CSR

This sub-section specifically focuses on pushing code from a local repository of the Cloud Shell Editor into CSR. A terminal can be opened within Cloud Shell to provide command-line instructions. A command-line approach to push code is elaborated on in the following procedure:

1. Switch to the console window in the Cloud Shell Editor.

2. Perform Git operations to add the new file and commit the changes:

   ```
   # Add new code file to the repository
   git add main.py


   # Commit the changes to the repository
   git commit -m "Adding main.py"
   ```

3. Push the local repository created in the Cloud Shell Editor with the new changes into the repository hosted in CSR. Indicate the appropriate project (after /p) and the destination repository in CSR (after /r):

   ```
   # Add local repository in Cloud Shell Editor as remote
   git remote add google \
   https://source.developers.google.com/p/gcp-devops-2021/r/
   my-first-csr
   # The above will create a remote repository with changes
   from the local repository
   # Push code to Cloud Source Repositories
   git push --all google
   # The above git push command will push the changes in the
   remote repository with specific project and repository
   name to google cloud source repositories
   ```

4. Navigate to the target repository (for example, my-first-csr) in CSR to view the newly added Python file, main.py.

Once the code is pushed to CSR from the remote branch, the code will be available in the master and is now ready to be deployed into any compute option of choice. The next topic illustrates the steps involved to download source code from CSR and deploy it into GCP's serverless compute option: Cloud Functions.

Creating a cloud function and deploying code from the repository in CSR

This sub-section specifically illustrates how CSR can integrate with other GCP compute options such as Cloud Functions:

1. Navigate to **Cloud Functions** in the GCP console.

2. Select the option to create a function (if a function is created for the very first time in a project, then this action will enable the Cloud Functions API).

3. Enter a function name of your choice and select a region, trigger type, and authentication mode. Save the options and continue. The following are examples:

    a) **Function name**: `my-first-csr`

    b) **Region**: `us-central1`

    c) **Trigger type**: `HTTP`

    d) **Authentication**: `Allow unauthenticated invocations`

4. Set the runtime as **Python 3.8** and the source code option as **Cloud Source Repository**.

5. Enter details related to the repository from which the code needs to be deployed and select the **Deploy** action (*Figure 5.19*):

    a) **Entry point**: Should be the name of the function that needs to be invoked by Cloud Functions. In this example, enter `hello_world` as the value.

    b) **Repository**: Should be the name of the repository in CSR where the source code is present. In this example, enter `my-first-csr` as the value.

    c) **Branch name**: Should be the branch where the code is present. In this example, enter `master` as the value.

    d) **Directory with source code**: Should be the directory path where the function mentioned as the entry point exists. In this example, enter / as the value:

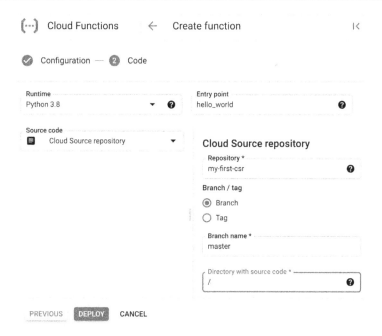

Figure 5.19 – Configuring the cloud source repository as the source code option

6. The function will be successfully deployed (*Figure 5.20*). The function can be tested either using the **Test Function** option under **Actions** in the list page or through the trigger URL specified under the **Details** section of the cloud function:

Figure 5.20 – Cloud function successfully deployed

This completes a detailed lab where the user makes a code change using GCP's Cloud Shell Editor, pushes to a repository using GCP's CSR, and deploys code to one of GCP's compute options such as Cloud Functions.

Summary

In this chapter, we discussed the service from Google Cloud to manage source code and provide Git version control to support collaborative development. This is the first key building block in the process of establishing a CI/CD process. In addition, we discussed various operations that can be performed in CSR along with a hands-on lab demonstrating native integration of CSR with Cloud Functions. The next chapter will focus on the services from Google Cloud required to build code, create image artifacts, and manage artifacts. These services include Cloud Build and Container Registry.

Points to remember

The following are some important points to remember:

- CSR is a fully managed private Git repository.

- CSR provides one-way sync with GitHub and Bitbucket.

- CSR provides a feature for universal code search and the search can be set either for a specific project, a specific repository, a specific directory, or everything.

- CSR can be set up to detect security keys. The currently supported types are service account credentials in JSON format and PEM-encoded private keys.

- CSR provides a feature to override security key detection at a commit level.

- Source Repository Reader/Writer/Admin are the supported access controls.

Further reading

For more information on GCP's Cloud Source Repositories, refer to the following:

- **Cloud Source Repositories**: `https://cloud.google.com/source-repositories`

Practice test

Answer the following questions:

1. Select the command that can be used to create a new repository in CSR called `my-first-csr` through the CLI:

    a) `gcloud create source repos my-first-csr`

    b) `gcloud source repos create my-first-csr`

    c) `gcloud create source repos my-first-csr`

    d) `gcloud source repo create my-first-csr`

2. Which of the following options allows one-way sync with CSR?

    a) GitHub

    b) Bitbucket

    c) None of the above

    d) Options *a* and *b*

3. Select the frequency of one-way sync from supported repository types to CSR:

    a) Every 5 minutes

    b) Configurable

    c) Real-time

    d) Near real-time

4. Which of the following is not a supported search filter in CSR?

    a) Search file contents

    b) Search by language

    c) Search by function

    d) Search by including terms

5. If the **Detect Security Keys** CSR feature is enabled, then when will CSR enforce this check?

    a) When a file is created

    b) `git add`

    c) `git push`

    d) `git commit`

6. Select the command to override security key detection at a commit level:

    a) `git push -o keycheck`

    b) `git push -o nokeycheck`

    c) `git push -o anykeycheck`

    d) `git push -o nonekeycheck`

7. Select two commands to enable and disable security key detection:

    a) `gcloud source project-configs update --enable-codeblock`

    b) `gcloud source project-configs update --enable-pushblock`

    c) `gcloud source project-configs update --disable-codeblock`

    d) `gcloud source project-configs update --disable-pushblock`

8. Which out of the following is not a valid access control with respect to CSR?

    a) Source Repository Reader

    b) Source Repository Writer

    c) Source Repository Editor

    d) Source Repository Admin

9. Which out of the following access controls can update a repository but cannot create one?

    a) Source Repository Reader

    b) Source Repository Writer

    c) Source Repository Editor

    d) Source Repository Admin

10. Which out of the following access controls can update repository configurations?

    a) Source Repository Reader

    b) Source Repository Writer

    c) Source Repository Editor

    d) Source Repository Admin

Answers

1. (b) – `gcloud source repos create "my-first-csr"`

2. (d) – Options *a* and *b*

3. (d) – Near real-time

4. (d) – Search by including terms

5. (d) – `git commit`

6. (b) – `git push -o nokeycheck`

7. (b) – `gcloud source project-configs update --enable-pushblock` and (d) – `gcloud source project-configs update --disable-pushblock`

8. (c) – Source Repository Editor

9. (b) – Source Repository Writer

10. (d) – Source Repository Admin

6
Building Code Using Cloud Build, and Pushing to Container Registry

The last chapter focused on managing source code using **Cloud Source Repositories** (**CSR**). CSR provides a fully managed private Git repository, provides one-way sync with GitHub and Bitbucket, and integrates with GCP services. This is the first step in the **Continuous Integration** (**CI**) flow.

This chapter will focus on the constructs required to build code, create image artifacts using Cloud Build and manage artifacts using GCP's Container Registry. This forms the crux of the CI workflow as the code is continuously built, artifacts are continuously created and stored in the registry, and application code is continuously deployed as containers.

In this chapter, we're going to cover the following main topics:

- **Key terminology** – Quick insights into the terminology around Docker and containers
- **Understanding the need for automation** – Understanding the need for automation by exploring the Docker life cycle

- **Building and creating container images** – Cloud Build essentials such as cloud builders and build configuration files, building code, storing and viewing build logs, managing access controls, and best practices to optimize the build speed

- **Managing container artifacts** – CSR essentials to push and pull images, manage access controls, configure authentication methods, and CI/CD integrations with CSR

- **Hands-on lab** – Step-by-step instructions to deploy an application to Cloud Run when a code change is pushed to the master branch

Technical requirements

There are three main technical requirements:

- A valid **Google Cloud Platform** (**GCP**) account to go hands-on with GCP services: `https://cloud.google.com/free`.

- Install Google Cloud SDK: `https://cloud.google.com/sdk/docs/quickstart`.

- Install Git: `https://git-scm.com/book/en/v2/Getting-Started-Installing-Git`.

Key terminology (prerequisites)

There are several key terminologies that are important to understand while trying to build, deploy, and maintain a distributed application that runs on containers. The following is a quick insight into some of those critical terminologies when dealing with containers:

- **Operating system** – An **Operating System** (**OS**) is system software that is critical to control a computer's hardware and software requirements across multiple applications, such as memory, CPU, storage, and so on. The OS coordinates tasks to ensure each application gets what it needs to run successfully. The OS consists of a kernel and software. The kernel is responsible for interacting with the hardware and the software is responsible for running the UI, drivers, file managers, compilers, and so on.

- **Virtualization** – Virtualization is the act of doing more with less by creating a virtual or software-based version of compute, storage, a network, and so on. It allows you to run multiple applications on the same physical hardware. Each application and its associated OS can run on a separate, completely isolated, software-based machine called a **virtual machine** or **VM**.

- **Hypervisor** – A hypervisor is software that creates and runs VMs, and essentially implements the concept of virtualization. A hypervisor allows one host computer to support multiple guest VMs by virtually sharing resources such as memory, storage, processing, and so on, and is responsible for giving every VM the required resources for peak performance.

- **Container** – A container is a unit of software that packages code and all its dependencies, which include libraries and configuration files. This enables applications to run quickly and reliably across computing environments. Containers use low-level OS constructs that allow you to specify unique system users, hostnames, IP addresses, filesystem segments, RAM, and CPU quotas.

- **Docker** – Docker is an open source platform for developing, building, deploying, and managing containerized applications. Docker uses OS-level virtualization to deploy or deliver software in packages called containers, providing the flexibility to run anywhere. Docker can also run any flavor of OS if the underlying OS kernel is Linux. As an example, containers can run different flavors of the Linux OS, such as Debian, CentOS, Fedora, and so on.

- **Docker daemon** – The Docker daemon represents the server that runs one or more containers. It is the service that runs the host OS. Additionally, the CLI represents the client, and the combination with the Docker daemon forms a client-server architecture.

- **Dockerfile** – A Dockerfile is a text document that contains a series or list of commands that can be executed from a command line in order to potentially assemble an image. A Dockerfile is the input for Docker to build images. The process automates the execution of a series of instructions or commands.

- **Docker layers** – A Docker layer represents an intermediate image that is created by executing each instruction in a Dockerfile. The link between the instruction and the intermediate image is stored in the build cache. A Docker container is essentially an image that has a readable/writable Docker layer built on top of multiple read-only images.

- **Docker images** – A Docker image consists of multiple Docker layers that are used to execute code in one or more containers. Essentially, a Docker image represents a plan that needs to be executed or, in other words, deployed.

The next section illustrates the Docker life cycle and emphasizes one of the key **Site Reliability Engineering** (**SRE**) objectives, which is to eliminate toil by investing in automation.

Understanding the need for automation

Once code is checked into a source code repository, the next step in a CI process is to build code and create artifacts as per the requirements to run the application. Once the artifacts are created, the artifacts are further stored in a repository and are later used by the **Continuous Deployment/Delivery (CD)** process to run the application. Given that the running theme in this book is to work with containers, Docker forms a key role as the OS-level virtualization platform to deploy applications in containers.

Following is an illustration of the Docker life cycle that highlights the multiple steps involved in creating container images to actually deploy containers that run the actual application:

1. The developer hosts code in a source code repository. The code can be changed during the development or enhancement process.

2. The source code repository can be set up to have trigger points, such as raising a pull request or merging code into a specific branch. The trigger points can be tied to initiate the code build process.

3. The code build process will look for a Dockerfile, which is essentially a set of instructions to create an application along with its dependencies.

4. A Dockerfile is used to create the build artifact – the container image, using `docker build`.

5. The created image can be pushed to an artifact repository to store container images, such as Docker Hub or GCP's Container Registry, and so on.

6. The application is created by downloading the container image from the repository into a compute environment and subsequently building a container, which essentially is a package that contains code, libraries, and dependencies.

If the preceding steps are converted into actual commands, then it will look like the following snippet:

```
#Build code using the Dockerfile
docker build -t <image-name> .


#Tag the locally created image with the destination repository
docker tag <image-name> <host-name>/<project-id>/<image-name>


#Push the tagged image to the choice of repository
docker push <host-name>/<project-id>/<image-name>
```

```
# Note that hostname refers to the location where image is
stored. 'gcr.io' refers that by default the images are stored
in Cloud Storage; specifically US location
#To deploy the application, pull the image from the repository
as a pre-requisite
docker pull <host-name>/<project-id>/<image-name>
```

```
#To deploy or run the application
docker run -name <container-name> <host-name>/<project-
id>/<image-name>
```

The steps mentioned as part of the Docker workflow are steps that need to be executed and in sequence. If there is a code fix or an incremental code change, then the steps need to be repeated in order to build, push, and deploy the code. This forms a repetitive or even an infinite loop, causing a lot of pain and suffering for developers. This is because the more manual steps there are, the greater the chance of human error. This qualifies as toil, since the steps are manual, repetitive in nature, devoid of any value, and can be automated.

Given that SRE's objective is to eliminate toil through automation, this forms a feasible approach to eliminate the infinite loop of pain and suffering. In addition, the preceding steps need to be executed in an environment that would need special attention or setup. For example, Docker will need to be set up to execute the preceding commands. In addition, the machine needs to have enough computing power and storage requirements to run the preceding steps in a repeated fashion. The machine also needs to be scaled if there are multiple parallel builds that are initiated at once.

GCP offers a service called **Cloud Build**, an automation engine that plays a key part in the CI/CD workflow. Cloud Build can import the source code, build in a managed workspace, and create artifacts such as Docker images, Java packages, binaries, and so on. Cloud Build can practically combine the steps to build, tag, and push a container image into a single configuration file. The container artifacts created by Cloud Build can be pushed and stored in another GCP service called **Container Registry**. The container image can be pulled from Container Registry at the time of container deployment. CloudBuild is capable of automating all these steps into a declarative syntax; also known as the build configuration file, which can be effectively run as many times as needed.

The upcoming sections will go into the details of the following:

- Cloud Build as the GCP service to build and create container images
- Container Registry as the GCP service to manage container artifacts

Building and creating container images – Cloud Build

Cloud Build is a service to build and create artifacts based on the commits made to source code repositories. The artifacts produced by Cloud Build can either be container or non-container artifacts. Cloud Build can integrate with GCP's CSR as well as popular external repositories such as GitHub and Bitbucket. Key features of Cloud Build include the following:

- **Serverless platform**: Cloud Build removes the need to pre-provision servers or pay in advance for computing power or storage required to build the code and produce artifacts. Based on the number of commits being made in parallel, scaling up or scaling down is an inherent process and doesn't require manual intervention.

- **Access to builder images**: Cloud Build provides cloud builders, which are pre-baked ready-to-use container images with support for multiple common languages and tools installed. For example, Docker Cloud Builders run the Docker tool.

- **The ability to add custom build steps**: Cloud Build requires a build config file where the list of steps can be explicitly specified by the user. The user can also specify the order of execution and include any dependencies as needed.

- **A focus on security**: Cloud Build supports vulnerability scanning and provides the ability to define policies that can block the deployment of vulnerable images.

The foundation for these Cloud Build features is based upon certain key elements that will be discussed in the upcoming sub-sections.

Cloud Build essentials

There are two key essential concepts with respect to Cloud Build, cloud builders and the build configuration.

Cloud builders

Cloud builders are container images that run the build process. The build process within a cloud builder is essentially a set of pre-defined build steps. In addition, a cloud builder can also include custom build steps. Cloud builder images are packaged with common languages and tools. Cloud Build can be used to run specific commands inside the builder containers within the context of cloud builders. Cloud builders can either be Google-managed, community-contributed, or public Docker Hub images.

Google-managed builders

Google provides managed pre-built images that can be used to execute one or more build steps. These pre-built images are in Google's Container Registry. Popular examples include docker builder (to perform `docker build`, `docker tag`, and `docker push` commands), gcloud builder (to perform the `docker run` command to deploy against a Google service such as Cloud Run), gke-deploy builder (to deploy in a GKE cluster), and so on. The complete list of Google-managed builders can be found at `https://github.com/GoogleCloudPlatform/cloud-builders`.

Community-contributed builders

Community-contributed builders are open source builders and are managed by the Cloud Build developer community. These are not pre-built images and, instead, only source code is made available by the developer community. Individual adaptations should build the source code and create an image. Popular examples include Helm (to manage the Kubernetes package), Packer (to automate the creation of images), and so on. The complete list of community-contributed builders can be found at `https://github.com/GoogleCloudPlatform/cloud-builders-community`.

Public Docker Hub builders

Public Docker Hub builders refers to publicly available Docker images that can be used to execute a set of build tasks. From a thought process standpoint, these builders are very similar to Google-managed builders but the images are not stored in Google Container Registry and are instead stored in Docker Hub. The complete list of public Docker Hub builders can be found at `https://hub.docker.com/search?q=&type=image`.

The build configuration

The build configuration is a configuration file that encapsulates the steps to perform build-related tasks. A build configuration file can be written in JSON or YAML format. The configuration steps specifically make use of cloud builders, which are either pre-built images (Google-managed or public Docker images) or images built by code maintained by the developer community, and essentially represent templated steps that could be reused with an option to pass explicit arguments. These templated steps can be used to fetch dependencies, perform unit and integration tests, and create artifacts using build tools such as Docker, Gradle, Maven, Bazel, and Gulp. An example of a build config file can contain instructions to build, package, and push Docker images to a container registry of choice. The structure of such a file will be detailed in the next sub-section.

Structure

A build config file consists of various fields or options. The most important of them is the build step (refer to *Figure 6.1*). There could be one or more build steps defined to reflect tasks required for the build process. Each build step essentially executes a Docker container and provides the flexibility to include multiple options:

- **Name**: Specifies a cloud builder that is a container image running common tools.

- **args**: Takes a list of arguments and passes it to the builder as input. If the builder used in the build step has an entry point, `args` will be used as arguments to that entry point; otherwise, the first element in `args` will be used as the entry point, and the remainder will be used as arguments.

- **Env**: Takes a list of environment variables in the form of a key-value pair.

- **dir**: Used to set a specific working directory. Optionally, artifacts produced by one step can be passed as input to the next step by persisting the artifacts in a specific directory. The directory path can either be a relative path (relative to the default working directory, which is `/workspace`) or a specific absolute path.

- **id**: Used to set a unique identifier for a build step.

- **waitFor**: Used if a specific build step is required to run prior. If not specified, then all prior steps need to be completed prior to the current build step.

- **entrypoint**: Used to override the default entry point provided by the cloud builder.

- **secretEnv**: Allows you to define a list of environment variables encrypted by Cloud KMS.

- **volumes**: Represents a Docker container volume that is mounted into build steps to persist artifacts across build steps.

- **timeout**: To specify the amount of time that a build can run. The default value is 10 minutes and the maximum allowed is 24 hours. Time should be specified in seconds.

Figure 6.1 shows the skeleton structure of a build configuration file that could consist of one or more build steps:

```
steps:
- name: string
  args: [string, string, ...]
  env: [string, string, ...]
  dir: string
  id: string
  waitFor: [string, string, ...]
  entrypoint: string
  secretEnv: string
  volumes: object(Volume)
  timeout: string (Duration format)
- name: string
  ...
- name: string
  ...
```

Figure 6.1 – Build steps in a build configuration file

Apart from the options that form the build steps of a build configuration file, additional possible options along with their details can be found at `https://cloud.google.com/cloud-build/docs/build-config`.

Building code using Cloud Build

The combination of cloud builders and build configuration files forms the core of Cloud Build. When Cloud Build is initiated, the following steps happen in the background:

1. The application code, Dockerfile, and other assets in a given directory are compressed.

2. The compressed code is then uploaded to a Cloud Storage bucket, which is either the default bucket created by Cloud Build on a per-project basis or a user-supplied Cloud Storage bucket.

3. A build is initiated with the uploaded files as input and the output of the build is a container image that is tagged with the provided image name.

4. The container image is then pushed to Container Registry or a destination registry of choice.

There are multiple approaches to invoke the build process via Cloud Build manual invocation and automatic builds using triggers.

Cloud Build – manual invocation via the gcloud CLI

There are two ways to initiate a build manually through Cloud Build using the `gcloud` command-line tool, which essentially uses the Cloud Build API:

- Using a Dockerfile
- Using Cloud Build – build configuration file

The upcoming sub-sections go into the details of the preceding two ways to initiate a build through Cloud Build.

Cloud Build – a manual build using a Dockerfile

The Dockerfile should contain all the information required to build a Docker image using Cloud Build. The following command will initiate the build process manually. This command should be run from the directory that contains the application code, Dockerfile, and any other required assets:

```
# Format to invoke the build manually using Dockerfile
gcloud builds submit --tag <host-name>/<project-id>/<image-
name> <app-code-directory-path>


#Example (. Indicates current directory)
gcloud builds submit –tag gcr.io/gcp-devops-2021/manual-
dockerfile-image .
```

Once the build is complete, the build ID will be displayed on the terminal or shell from where the `build` command was invoked. The build ID can be used to filter through the builds displayed in the Cloud Build console and is subsequently useful to view the build logs. Additionally, the newly created image will be pushed to Container Registry as per the preceding example.

Cloud Build – a manual build using a build configuration file

Another approach to initiate a manual build through Cloud Build is to use a build configuration file. The build configuration file uses cloud builders, which essentially are critical to minimize the manual steps in a templated specification file.

The following is an example build configuration file that uses docker cloud builder to build code and push an image to Container Registry. The name of the container image used here is `builder-myimage` and the name of the configuration file is `cloudbuild.yaml`:

```
steps:
- name: 'gcr.io/cloud-builders/docker'
  args: ['build', '-t', 'gcr.io/$PROJECT_ID/builder-myimage',
'.']
- name: 'gcr.io/cloud-builders/docker'
  args: ['push', 'gcr.io/$PROJECT_ID/builder-myimage']
- name: 'gcr.io/cloud-builders/gcloud'
```

The following command will initiate the Cloud Build process by using the build configuration file (which is `cloudbuild.yaml` in this case) as the input, along with the path to the source code:

```
# Format to invoke the build manually using the build
configuration file
gcloud builds submit --config <build-config-file> <source-code-
path>


#Example 1 (Source code is located in the current directory)
gcloud builds submit --config cloudbuild.yaml .


#Example 2 (Source code is located in a cloud storage bucket)
gcloud builds submit --config cloudbuild.yaml gs://my-cloud-
build-examples/cloud-build-manual.tar.gz
```

Cloud Build – automatic build using triggers

The manual invocation of Cloud Build does not fit into the CI/CD workflow as it adds toil. The preferred approach is to automatically build code whenever a qualified event is detected. Cloud Build facilitates this feature by using the option of triggers.

The user can create a trigger that could be invoked on one of the following qualifying events:

- Push to a branch.

- Push a new tag.

- A pull request (GitHub app only).

The trigger continuously monitors for the configured event against the configured repository. If the event occurs, the trigger initiates the build process using either the Dockerfile or Cloud Build configuration file (as configured on the trigger) and subsequently, the build process will result in build artifacts. A step-by-step hands-on lab is illustrated toward the end of this chapter.

> **Dockerfile versus cloudbuild.yaml**
>
> A Dockerfile allows you to build and compose a Docker container image using the `docker build` command. A Dockerfile also allows you to incorporate build steps using bash commands; they could include commands specific to Google Cloud; after specifying the installation of Google Cloud SDK as one of the steps.
>
> On the contrary to using a Dockerfile, `Cloudbuild.yaml` also allows you to build and compose a Docker container image and to utilize Google-managed or community-managed builders that come with pre-built images and offer more customization. The choice between the two comes to intent, choice of cloud platform, and ease of customization.

This concludes the sub-section on how a build can be initiated through Cloud Build. The next sub-section focuses on details related to storing and viewing build logs.

Storing and viewing build logs

Cloud Build creates a log trail for actions performed as part of a build process. This log information is stored in Cloud Logging. Additionally, Cloud Build stores the log information in a Cloud Storage bucket. In fact, a default Cloud Storage bucket is created on a per-project basis, once the Cloud Build API is enabled. The bucket is named in the format `<project-id_cloudbuild>`. The logs related to every build are compressed and stored in the storage bucket.

So, the default option to store the Cloud Build logs is both in Cloud Logging as well as a Cloud Storage bucket. However, it is possible to choose either of the two options specifically in the build configuration file by using the *logging* field:

- If set to CLOUD_LOGGING_ONLY, then logs are written only to Cloud Logging.
- If set to GCS_ONLY, then logs are written only to the Cloud Storage bucket. The default bucket will be used unless a Cloud Storage bucket is explicitly specified using the **logsBucket** option.

It's possible that the user will go with an option other than the default options either due to cost constraints or it's possible that logs are ingested to another logging framework with the Cloud Storage bucket being the source.

The following is a code snippet that demonstrates the usage of the *logging* option as part of the build configuration file:

```
steps:
- name: 'gcr.io/cloud-builders/docker'
  args: ['build', '-t', 'gcr.io/myproject/myimage', '.']
options:
  logging: GCS_ONLY
logsBucket: 'gs://mylogsbucket'
```

Logs can be viewed using the Cloud Logging console. If logs need to be viewed at an individual build level, it is preferred to view the logs from the Cloud Build console. The information in the Cloud Build console will be derived from the Cloud Storage bucket (either the default or the explicit bucket). In order to view the logs, the user should either have the *Storage Object Viewer* role or the *Project Viewer* role.

To view the build logs, follow these steps:

1. Navigate to **Cloud Build** in the GCP Console (by default, the user will be taken to the **Build History** page).
2. Select a build to view its respective logs (builds that succeeded will be in green, and in red otherwise).

3. The user can view the build log per build step. In addition, execution details and the storage locations of any relevant build artifacts are also displayed. Optionally, the source of the cloud logs is also shown (refer to *Figure 6.2*):

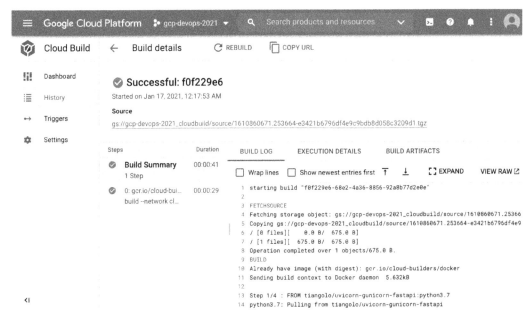

Figure 6.2 – Build log from Cloud Build

If a need arises to delete the build logs, then logs cannot be deleted from a Google-created log bucket. However, logs can be deleted from a user-created log bucket or by deleting the user-created bucket itself that contains one or more build logs. This requires the user to have access to Cloud Storage to delete a file – through Cloud Storage; specifically, the role Storage Admin or Storage Object Admin (depending upon whether the intention is to delete the entire user-created bucket or the specific build log file respectively).

Managing access controls

A build can be triggered either by a user or by an application. As per Google's recommended practices, if an application needs access to a service, then it can be possible through a service account. So, to be precise, access control to Cloud Build can either be managed via **end user IAM roles** or through a **Cloud Build service account**.

End user IAM roles

Cloud Build has a set of predefined IAM roles that can provide granular access and can also align to a specific job role. This prevents unwanted access and allows you to implement the principle of least privilege.

The following table summarizes the critical IAM roles required to access or perform actions on Cloud Build:

Role Name	Role Description
Cloud Build Viewer	Can view Cloud Build resources. Allows you to get or list builds and associated triggers.
Cloud Build Editor	Full control of Cloud Build resources. In addition to Cloud Build Viewer, allows you to create a build, cancel a build, and create/patch/delete/run a trigger.

Cloud Build service accounts

Google recommends using a **service account (SA)** when a task needs to be performed by an application or on behalf of a user. A service account is a special kind of account that is used by an application or a VM to make authorized API calls but not by an individual. The regular practice in such scenarios is to create a SA and assign the necessary permissions to the SA so that the application with that SA can perform the necessary actions.

Cloud Build instead creates a specific Cloud Build SA for a project when the Cloud Build API is enabled on the project. The Cloud Build SA has a minimal number of permissions assigned to it, for example, Cloud Storage. If you want to use other services, the SA needs to be updated to reflect the desired permissions.

The set of pre-assigned permissions for the Cloud Build SA will essentially allow Cloud Build to perform the following actions on behalf of the user:

- Create, list, get, or cancel builds.
- Create, patch, delete, or run a build trigger.
- Pull source code from CSR.
- Store images in and get images from Container Registry.
- Store artifacts in and get artifacts from Cloud Storage.
- Store artifacts in and get artifacts from Artifact Registry.

- Create build logs in Cloud Logging.

- Store build logs in a user-created logs bucket.

- Push build updates to Pub/Sub.

- Get project information and list projects.

This concludes the topic on managing access controls, giving insights into the required IAM roles. The upcoming topic focuses on best practices while executing the build process, which could essentially reduce the build execution time.

Cloud Build best practices – optimizing builds

Decreasing the build time helps in optimizing the build process. Given that the focus is on handling containers, there are two common strategies to increase the build speed:

- **Building Leaner Containers**: As a part of this strategy, the size of a container can be reduced if files related to build-time dependencies and any intermediate files are not included in the container image.

- **Cached Docker images**: As a part of this strategy, a cached image can be specified via the `--cache-from` argument and can be used for subsequent builds as the starting point. The cached image will be retrieved from a registry. A cached Docker image is only supported for Docker builds and is not supported by cloud builders.

In addition to a generic strategy of building leaner containers to optimize the build speed, Cloud Build specifically prescribes the following best practices, which can additionally be used:

- Kaniko cache

- Cloud Storage for caching directories

- Custom VM sizes

- Ignoring unwanted files

The following are the details of the above-mentioned best practices.

Kaniko cache

Kaniko cache is based on the open source tool Kaniko and is also a feature of Cloud Build where intermediate container image layers are directly written to Google's Container Registry without an explicit push step.

To enable Kaniko cache, as part of the build configuration file `cloudbuild.yaml`, the following is a code snippet that could incorporate it:

```
steps:
- name: 'gcr.io/kaniko-project/executor:latest'
  args:
  - --destination=gcr.io/$PROJECT_ID/image
  - --cache=true
  - --cache-ttl=XXh
```

The following are recommendations that should be taken into consideration while implementing Kaniko cache through the `kaniko-project` cloud builder:

- Use `kaniko-project/executor` instead of `cloud-builders/docker`.
- The `destination` flag should refer to the target container image.
- The `cache` flag should be set to `true`.
- The cache-ttl flag should be set to the required cache expiration time.

Alternatively, Kaniko cache can be enabled via the gcloud CLI by running the command as shown in the following snippet:

```
gcloud config set builds/use_kaniko True
```

Kaniko cache speeds up the build execution time by storing and indexing the intermediate layers within a Container Registry and eventually saves build execution time since it can be used by subsequent builds.

Cloud Storage for caching directories

Conceptually, this is like a cached Docker image. The results of a previous build can be reused by copying from a Cloud Storage bucket and the new results can also be written back to the Cloud Storage bucket. This concept is not restricted only to Docker builds but can also be extended to any builder supported by Cloud Build.

Additionally, Cloud Build uses a default working directory named `/workspace`, which is available to all steps in the build process. The results of one step can be passed on to the next step by persisting it in the default working directory. The working directory can also be explicitly set using the `dir` field as part of the build step.

The following is a sample snippet of a build configuration file where Cloud Storage is used for caching directories:

```
steps:
- name: gcr.io/cloud-builders/gsutil
  args: ['cp','gs://mybucket/results.zip','previous_results.
zip']
  dir: 'my-cloud-build/examples'
# operations that use previous_results.zip and produce new_
results.zip
- name: gcr.io/cloud-builders/gsutil
  args: ['cp','new_results.zip','gs://mybucket/results.zip']
  dir: 'my-cloud-build/examples'
```

The preceding example also shows the usage of a specific working directory, `my-cloud-build/examples`, as specified under the `dir` field as part of the build steps. Like Kaniko cache, cloud storage can be used to optimize build speeds by using the results from a previous build.

Custom VM sizes

Cloud builds are executed against a managed VM of a standard size. However, Cloud Build provides an option to increase the speed of a build by using a higher CPU VM, which essentially speeds up the build process. This is done by specifying the `--machine-type` argument. Cloud Build specifically provides a choice of 8 cores or 32 cores across two families of VMs. Specific choices are as follows:

- N1_HIGHCPU_8

- N1_HIGHCPU_32

- E2_HIGHCPU_8

- E2_HIGHCPU_32

The following is the CLI command to specify a machine type while initiating the Cloud Build process:

```
gcloud builds submit --config=cloudbuild.yaml \
  --machine-type=N1_HIGHCPU_8
```

Ignoring unwanted files

Cloud Build uploads the code directory to a Cloud Storage location. The upload process can be made quicker by ignoring files that are not relevant to the build process. These files might include third-party dependencies, compiled code, binaries, or JAR files used for local development. In addition, documentation and code samples are not required for the build process. These files can be specified as part of the `gcloudignore` file to optimize the upload time.

This completes our deep dive into Cloud Build and its key constructs, which include cloud builders and the build configuration, options available to initiate a build process, automating the available options using triggers, viewing build results with information stored in Cloud Storage, defining access controls, and prescribing recommended practices to optimize builds.

The next section focuses on the concepts of artifact management and the usage of Container Registry to manage build artifacts while working with containers.

Managing build artifacts – Container Registry

Source code management is the first step in the CI process. This is followed by building the code. Code can be built based on various trigger points; either against a development branch or when a PR is merged into the master branch. The code build process can result in one or more artifacts. Based on the nature of the code being built, the resultant artifacts can either be binaries, packages, container images, or a combination. These artifacts are stored in a registry and then deployed into a computing environment and form the CD process. In between the CI and CD process, there is an intermediate process where the build artifacts are stored and then subsequently deployed. This is known as **artifact management**.

Artifact management acts as a single source of truth and a critical integration point between CI and CD. Many artifact management systems provide versioning, the ability to scan for vulnerabilities, provide consistent configuration, and accommodate unified access control.

Given that the theme of this book is working with containers, the critical build artifacts in this case will be the container images. Images are typically stored in a central registry. The most common container registry is Docker Hub, which stores public Docker images. However, when working within an enterprise, it is generally a requirement to secure access to the container images produced by building code that is specific to the enterprise. In such scenarios, a private registry is preferred over a public registry, since a private registry can offer role-based access controls to provide more security and governance.

Container Registry is GCP's private container image registry service, which supports Docker Image Manifest V2 and OCI image formats including Docker. The Container Registry service can be accessed through secure HTTPS endpoints and allows users to push or pull images from any possible compute option.

> **Artifact Registry**
>
> Artifact Registry is a managed service offering from GCP that is similar to Container Registry but also provides options to store non-container artifacts such as Java packages, Node.js modules, and so on. It is currently not part of the GCP DevOps Professional exam.

Container Registry – key concepts

Container Registry is one of Google's approaches to artifact management. Like any other service, it has certain key constructs and concepts. The following sub-sections dive into those details.

Enabling/disabling Container Registry

The Container Registry service can be enabled or disabled using the GCP Console via the **APIs & Services** section. Additionally, the service can be enabled or disabled through the CLI using the following command:

```
# To enable container registry
gcloud services enable containerregistry.googleapis.com
# To disable container registry
gcloud services disable containerregistry.googleapis.com
```

Container Registry service accounts

Like Cloud Build, when Container Registry is enabled, a Google-managed SA will get created that is specific to your current project. This SA allows Container Registry to access critical GCP services such as Pub/Sub and Cloud Storage within the project. Google makes this possible by assigning the Container Registry Service Agent role to the Container Registry SA.

The structure of Container Registry

There could be one or more registries in a Container Registry service. Each registry is identified by the hostname, project ID, and image (tag or image digest). The following are the two possible formats:

- `HOSTNAME / PROJECT_ID / IMAGE:TAG`
- `HOSTNAME / PROJECT_ID / IMAGE@IMAGE-DIGEST`

In the preceding code, we have the following:

- `HOSTNAME`: Refers to the location where the image is stored. Images are stored in a Cloud Storage bucket. If the hostname is `gcr.io`, then by default the images are stored in the United States. Additionally, the user can specify specific hosts such as `us.gcr.io`, `eu.gcr.io`, or `asia.gcr.io`, where each host is tied to a specific geographic region where the images are hosted.
- `PROJECT_ID`: Refers to the specific GCP project ID.
- `IMAGE`: Refers to the image name. Registries in Container Registry are listed by image name. A single registry can hold different versions of an image. Adding either `:TAG` or `@IMAGE-DIGEST` helps to differentiate between images with the same image name. If neither is specified, then the image is tagged as latest.

Examples:

The following are examples of a registry for a specific image where the version of the image is differentiated by either adding a tag or image digest:

```
# Add image tag:
gcr.io/PROJECT-ID/my-image:tag1
```

```
# Add image digest:
gcr.io/PROJECT-ID/my-image@sha256:4d11e24ba8a615cc85a535daa17
b47d3c0219f7eeb2b8208896704ad7f88ae2d
```

This completes the topic that details the structure of Container Registry, an understanding that is critical to upload or download container images to/from Container Registry. This will be detailed in upcoming topics.

Uploading images to Container Registry

The build process, on completion, will produce container images as artifacts. These artifacts are generally created in the local directory where the build process was run. These local Docker images need to be uploaded to a private registry such as Container Registry. The process of uploading an image to Container Registry is also synonymous with pushing images to Container Registry.

To break it down, there are two main steps that push images to Container Registry:

1. Tag the local image with the registry name (as shown in the following snippet):

    ```
    docker tag SOURCE_IMAGE HOSTNAME/PROJECT_ID/IMAGE
    #Example
    docker tag my-local-image gcr.io/gcpdevops-2021/my-gcr-image
    ```

2. Push the tagged image to Container Registry (as shown in the following snippet):

    ```
    docker push HOSTNAME/PROJECT-ID/IMAGE
    #Example
    docker push gcr.io/gcpdevops-2021/my-gcr-image
    ```

A container image can be pushed to a new registry or an existing registry:

- If pushed to a new registry, that is, a registry with a new hostname, then Container Registry will create a multi-regional storage bucket.

- If pushed to an existing registry, then a new version of the image is created either with an image tag or image digest. If neither is present, then the image is tagged as `latest`.

> **Specifying the location of Container Registry**
>
> The location of Container Registry can be specified under the hostname. If `gcr.io` is used, then the default location is *United States*. If a specific location needs to be used, then the host can be specified as `eu.gcr.io`.

The newly created image can be listed using the following gcloud CLI command:

```
gcloud container images list –repository=HOSTNAME/PROJECT-ID
#Example
gcloud container images list –repository=gcr.io/gcpdevops-2021
```

This concludes the topic on uploading or pushing a container image to GCP's Container Registry. Now the newly pushed image can be deployed by any application by downloading the image from Container Registry. This will be covered as the next topic.

Downloading images from Container Registry

The CD process feeds on the output of the CI process, which essentially is stored in a registry such as Container Registry in the form of an OCI image. So, for the CD process to progress, the Docker image needs to be downloaded from Container Registry. The process of downloading an image from Container Registry is synonymous with pulling images from Container Registry.

An image can be pulled from Container Registry either using the image tag or image digest. If neither is specified, then the image with a tag of `latest` will be downloaded (as shown in the following snippet):

```
# Pull based on Image Tag
docker pull HOSTNAME/PROJECT-ID/IMAGE:TAG
# Pull based on Image-Digest
docker pull HOSTNAME/PROJECT-ID/IMAGE@IMAGE_DIGEST
# Pull without Image Tag or Image-Digest
docker pull HOSTNAME/PROJECT-ID/IMAGE
```

This completes the topic on downloading images from Container Registry. To either upload or download images to or from Container Registry, it is critical that the user or application trying to perform those actions has the necessary access controls. This will be covered as the next topic.

Container Registry access controls

Container Registry is a repository for container images. The images are physically stored in a Cloud Storage bucket. So, in order to push or pull images from Container Registry, the user or SA should be granted the following roles:

Role Name	Role Description
Storage Admin	Provides the ability to push and pull images from the Cloud Storage bucket associated with Container Registry
Storage Object Viewer	Provides the ability to pull images from the Cloud Storage bucket associated with Container Registry

If an application is deployed using GCP's available compute options, such as Compute Engine, App Engine, or GKE, then each of these services will have default service accounts with a pre-defined set of roles. However, the use of default service accounts is not recommended as this practice does not follow the principle of least privilege. Alternatively, it is also possible that the compute options could use a custom SA with the minimum set of required permissions. Either way, it is important to understand the scope of these service accounts and their impact during the CD process. This will be discussed in detail in the next topic.

Continuous Delivery/Deployment integrations via Container Registry

As mentioned previously, artifact management is the bridge between CI and CD. GCP has multiple compute options where code or an application can be deployed as part of the CD process. Each of GCP's compute options has a way to interact and integrate with Container Registry, which are detailed in the following sub-sections.

Compute Engine

The **Compute Engine** service uses either a SA or access scopes to identify the identity and provide API access to other services. The following is a summary of the possibilities or potential changes to successfully push or pull an image originating from a Compute Engine instance:

- The default Compute Engine SA or the default access scope provides read-only access to storage and service management. This allows you to download or pull images from Container Registry within the same project.

- To push images, either the read-write storage access scope should be used, or the default Compute Engine SA should be configured with the Storage Object Admin role.

- If the VM instance is using a SA other than the default Compute Engine SA or if the VM instance is in a project different from Container Registry, then the SA should be given the appropriate permissions to access the storage bucket used by Container Registry.

Google Kubernetes Engine

A **Google Kubernetes Engine** (**GKE**) cluster is essentially a collection of Google Compute Engine VMs that represents a node pool. This also means that GKE uses the SA configured on the VM instance. So, eventually, GKE's access to Container Registry is based on the access granted to the VM's SA. So, refer to the previous sub-section on *Compute Engine* for the possibilities or potential changes to successfully push or pull an image originating from a compute instance within GKE.

App Engine flexible

App Engine flexible supports the deployment of Docker containers. The default SA tied with App Engine flexible has the required permissions to push and pull images from Container Registry, provided both are present in the same project.

If App Engine is in a different project than Container Registry or if App Engine is using a different SA than the default App Engine SA, then the SA tied to App Engine should be given the appropriate permissions to access the storage bucket used by Container Registry.

This completes the topic on how GCP compute options can integrate with Container Registry. Outside the compute options provided by GCP, there are several use cases where CD systems use a compute option that is not native to GCP.

The next topic discusses the details of how third-party clients can access artifacts in GCP's Container Registry.

Container Registry authentication methods

There are compute options outside Google Cloud that could potentially deploy an application by pulling container images from Google Cloud's Container Registry. Such compute options are referred to as third-party clients. A **Red Hat Enterprise Linux** (**RHEL**) cluster is an example of a third-party client that is a compute option from Red Hat and can download a container image from Container Registry.

Apart from ensuring that the third-party client has the required access control to pull or push images, it is mandatory for third-party clients to authenticate with Container Registry prior to initiating an attempt to push or pull images. The following are the possible authentication methods that third-party clients can use to authenticate with Container Registry:

- gcloud credential helper
- Standalone credential helper

The details on how third-party clients can authenticate with Container Registry are elaborated on in the upcoming sub-sections.

gcloud credential helper

This is the recommended authentication method and mandates the installation of Google's Cloud SDK or the usage of GCP's Cloud Shell. This method essentially uses the gcloud tool to configure authentication. The following are the required steps to use this authentication method:

1. Log into gcloud as the IAM user that will run the Docker commands:

   ```
   # To configure authentication with IAM user credentials:
   gcloud auth login
   ```

2. If the intent is to log into gcloud as a SA, then run the following Docker command. This uses a JSON key file that contains the information about the SA and retrieves an access token that is valid for 60 minutes:

   ```
   # To configure authentication with service account
   credentials:
   gcloud auth activate-service-account <service-account-
   name> --key-file=<key-file-name>
   ```

3. Configure Docker with the following command, which allows Docker to authenticate with Container Registry:

   ```
   gcloud auth configure-docker
   ```

The next sub-section details an alternative approach to how the Docker standalone credential helper can be used as an authorization method for third-party clients to interact with GCP's Container Registry.

Standalone credential helper

`docker-credential-gcr` is GCP Container Registry's standalone credential helper. This authentication method is used when Google Cloud SDK is not installed or GCP Cloud Shell is not used but Docker needs to be configured to authenticate with GCP Container Registry. This credential helper implements the Docker Credential Store API and enables more advanced authentication schemes for using GCP Container Registry. It allows you to fetch credentials from application default credentials and is also capable of generating credentials without an explicit login operation. More details about `docker-credential-gcr` can be found at `https://github.com/GoogleCloudPlatform/docker-credential-gcr`.

The following are the required steps to use the standalone credential helper as the authentication method:

1. Log on to the machine as the user that will run the Docker commands.

2. Download `docker-credential-gcr` from GitHub releases: (`https://github.com/GoogleCloudPlatform/docker-credential-gcr/releases`).

3. Configure Docker with the following command. Internally, the credential helper will use a SA that is supplied in a JSON key file:

```
docker-credential-gcr configure-docker
```

Container Analysis

Container Analysis is a construct of Container Registry and even Artifact Registry. The purpose of this construct is to analyze the image that is being pushed into GCP Container Registry for any vulnerabilities that might be a security concern. The resulting metadata from vulnerability scanning is stored and is made available through an API for consumption. This metadata is later used in the authorization process.

There are two specific APIs through which Container Analysis provides metadata storage and vulnerability scanning:

- **Container Analysis API**: Enables metadata storage. Metadata storage includes information about vulnerability or build information, also referred to as *note*.

- **Container Scanning API**: Enables vulnerability scanning across the project. The process comprises scanning and continuous analysis to find malicious activity or potential compromises leading to system failure.

The following are the steps involved in configuring Container Analysis as part of Container Registry:

1. Enable the Container Analysis API: Navigate to **APIs & Services**, search for `Container Analysis API`, and select the **Enable** option.

2. Enable the Container Scanning API: Navigate to **APIs & Services**, search for `Container Scanning API`, and select the **Enable** option. In addition, also search for `On-Demand Scanning API` and enable it.

3. Navigate to **Container Registry** and under **Settings**, verify that **Vulnerability Scanning** is enabled. If enabled, the **Settings** screen will be similar to *Figure 6.3*. If not, enable it:

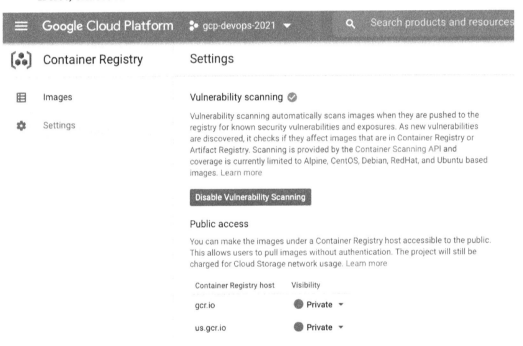

Figure 6.3 – Vulnerability scanning enabled in Container Registry

4. Now when an image is pushed to Container Registry, container analysis and vulnerability scanning will be performed automatically. The results will be displayed under the **Images** section of **Container Registry**. *Figure 6.4* represents the summary of the container analysis:

Figure 6.4 – Summary of container analysis on a newly created image

5. The details of all the vulnerabilities scanned and the categorization of them can be found by clicking on the summary. *Figure 6.5* represents the detailed report:

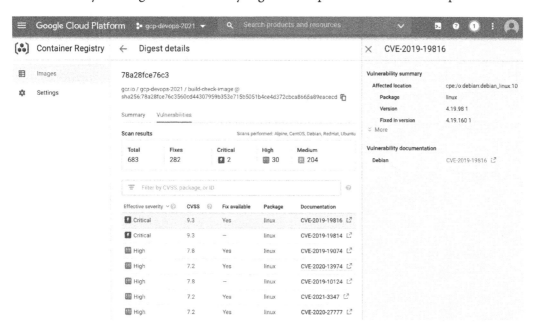

Figure 6.5 – Details of vulnerability scanning through Container Analysis

This completes multiple sub-sections related to how GCP's compute options, as well as other third-party CD systems, can integrate with GCP Container Registry. This also concludes the deep dive into several of the key factors related to Container Registry.

The next section is a hands-on lab that tries to combine multiple concepts learned across sections of this chapter.

Hands-on lab – building, creating, pushing, and deploying a container to Cloud Run using Cloud Build triggers

The goal of this hands-on lab is to provide a step-by-step illustration of how code can be automatically built, pushed, and deployed to a compute option called Cloud Run.

> **Cloud Run**
> Cloud Run is GCP's managed serverless compute option, which deploys containers and abstracts away the infrastructure management. Cloud Run can scale up or down from zero based on traffic and charges on a pay-per-use model.

The hands-on lab implements concepts across Cloud Build and Container Registry. The following is a high-level breakdown of the steps involved. Each of the steps is further elaborated into multiple sub-steps:

1. Creating an empty repository in Source Repositories
2. Creating a Cloud Build trigger
3. Adding code and pushing it to the master branch
4. Code walk-through to build, create, push, and deploy the container image
5. Viewing the build results in Cloud Build, Container Registry, and Cloud Run

Let's take a look at these steps in detail.

Creating an empty repository in Source Repositories

The following are the steps required to create an empty repository in GCP's Source Repositories:

1. Navigate to **Source Repositories** in the GCP Console.

2. Create a new repository by using the **Add repository** option under the **All repositories** tab. Select an appropriate project and name the repository as per your choice (for example, `my-cloud-build-trigger`).

Creating a Cloud Build trigger

The following are the steps required to create a Cloud Build trigger against a specific repository, which will be invoked on a specific repository event (refer to *Figure 6.6*):

1. Navigate to the **Triggers** section under **Cloud Build** in the GCP console.

2. Select the **Create Trigger** option.

3. Enter an appropriate name for the trigger, for example, `build-on-push-to-master`.

4. Enter an appropriate description.

5. Select a choice of event. Available options are **Push to a branch**, **Push new tag**, or **Pull request**. In this specific example, select the **Push to a branch** option.

6. Select a source repository. In this specific example, select the newly created repository, that is, `my-cloud-build-trigger`.

7. Select a choice of branch. It can be * or a specific branch. In this specific example, enter the option as `^master$`.

8. Select the source for the build configuration. It can either be a Cloud Build configuration file or a Dockerfile. In this specific example, select the **Cloud Build configuration file** option and accordingly provide the file location (as `/cloudbuild.yaml`).

9. Create the Cloud Build trigger (refer to *Figure 6.6*):

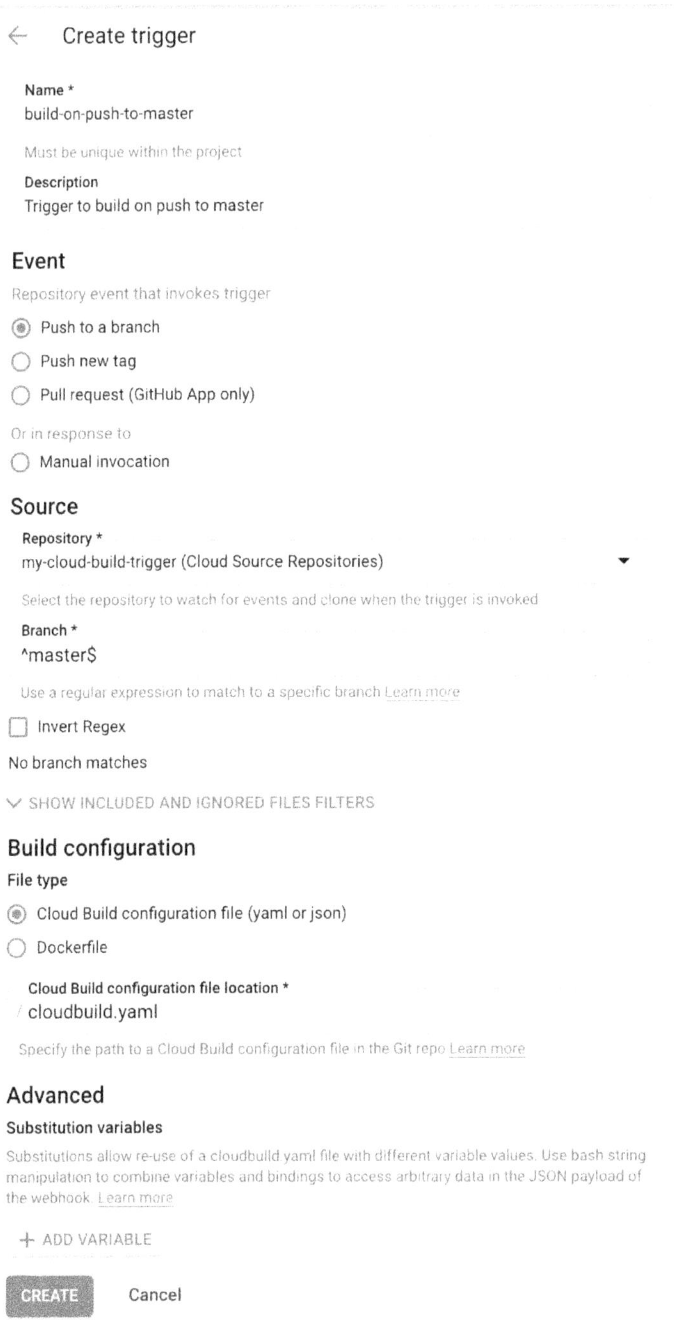

←   Create trigger

Name *
build-on-push-to-master

Must be unique within the project

Description
Trigger to build on push to master

Event

Repository event that invokes trigger

◉ Push to a branch

○ Push new tag

○ Pull request (GitHub App only)

Or in response to

○ Manual invocation

Source

Repository *
my-cloud-build-trigger (Cloud Source Repositories)   ▼

Select the repository to watch for events and clone when the trigger is invoked

Branch *
^master$

Use a regular expression to match to a specific branch Learn more

☐ Invert Regex

No branch matches

∨ SHOW INCLUDED AND IGNORED FILES FILTERS

Build configuration

File type

◉ Cloud Build configuration file (yaml or json)

○ Dockerfile

Cloud Build configuration file location *
/ cloudbuild.yaml

Specify the path to a Cloud Build configuration file in the Git repo Learn more

Advanced

Substitution variables

Substitutions allow re-use of a cloudbuild yaml file with different variable values. Use bash string manipulation to combine variables and bindings to access arbitrary data in the JSON payload of the webhook. Learn more

+ ADD VARIABLE

CREATE   Cancel

Figure 6.6 – Steps to illustrate the creation of the Cloud Build trigger

Adding code and pushing it to the master branch

We have created a repository and set up a trigger against the repository. The trigger will build code when the code is pushed to the master branch. The next step is to add code to the repository and push it to the master branch. The following steps illustrate this:

1. Clone the empty repository to a local Git repository:

   ```
   gcloud source repos clone my-cloud-build-trigger
   --project=$GOOGLE_CLOUD_PROJECT

   #$GOOGLE_CLOUD_PROJECT is an environment variable that
   refers to the current project
   ```

2. Switch to the new local Git repository:

   ```
   cd my-cloud-build-trigger
   ```

3. Create a remote branch:

   ```
   git checkout -b feature/build-trigger
   ```

4. Copy the my-cloud-build-trigger folder from https://github.com/ PacktPublishing/Google-Cloud-Platform-for-DevOps-Engineers/ tree/main/my-cloud-build-trigger.

5. Add files, commit the changes, and push to the remote branch:

   ```
   git add *
   git commit -m "Adding files!"
   git push --set-upstream origin feature/build-trigger
   ```

6. Checkout to the master branch and fix the upstream:

   ```
   git checkout -b master
   git branch --unset-upstream
   ```

7. Merge the remote branch with the master branch:

   ```
   git push –set-upstream origin git push -u master
   ```

Code walk-through

As soon as the code is pushed to the master branch in the previous step, the configured trigger will come into effect and will eventually build the code, create a container image, push the container image to Container Registry, and eventually provide the feasibility of the container image being deployed.

The `my-cloud-build-trigger` repository consists of three types of files:

- The application code
- Dockerfile
- The build configuration file

The application code

The application code represents the core code that runs the application. In this specific case, the code is under `app/main.py`, is written in Python, and creates a web application using the FastAPI framework. The following is the code snippet:

```
app = FastAPI()
@app.get("/")
def read_root():
    return {"Hello": "World"}
```

Dockerfile

The Dockerfile represents the instructions required to build the application code using a base image and subsequently create a container image. The following is the code snippet:

```
FROM tiangolo/uvicorn-gunicorn-fastapi:python3.7
COPY ./app /app
EXPOSE 8080
CMD ["uvicorn", "main:app", "--host", "0.0.0.0", "--port",
"8080"]
```

The build configuration file

The build configuration file represents the configuration to initiate the build process. In addition, it can include steps to push the container image to Container Registry and subsequently deploy it. The following is the code snippet:

```
steps:
- name: 'gcr.io/cloud-builders/docker'
```

```
    args: ['build', '-t', 'gcr.io/$PROJECT_ID/cloud-build-
trigger', '.']

- name: 'gcr.io/cloud-builders/docker'
  args: ['push', 'gcr.io/$PROJECT_ID/cloud-build-trigger']

- name: 'gcr.io/cloud-builders/gcloud'
  args: ['run', 'deploy', 'cbt-cloud-run', '--image', 'gcr.
io/$PROJECT_ID/cloud-build-trigger', '--region', 'us-central1',
'--platform', 'managed', '--allow-unauthenticated']
```

In this specific example, the configuration file has three specific steps:

1. Build the code using Docker Cloud Builder. The code is picked up from the specified directory. In this case, it is the current directory.

2. The code built in the first step creates a container image that is local to the cloud builder. The image is then tagged and pushed to Container Registry using the Docker Cloud Builder. The container image is pushed against a specific repository.

3. The image pushed in *step 2* is used in this step to deploy to Google's Cloud Run.

Viewing the results

After the code is pushed to the master branch, the configured trigger will initiate the build process. To view the build results, navigate to the **History** section of **Cloud Build** in the GCP console and find the build result for the specific source repository (refer to *Figure 6.7*):

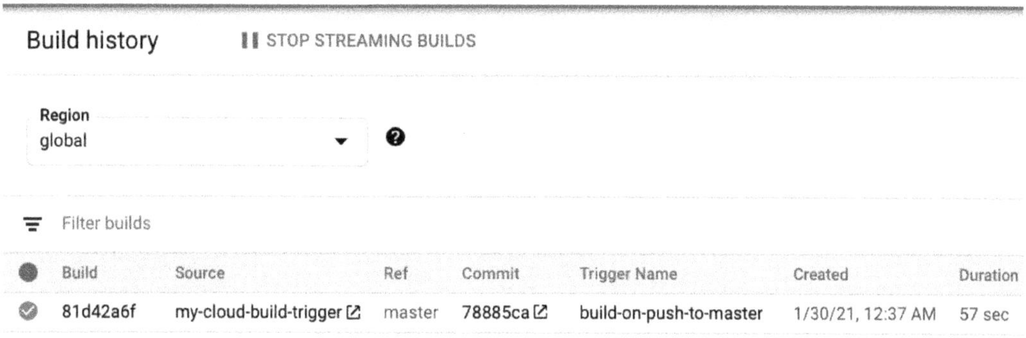

Build	Source	Ref	Commit	Trigger Name	Created	Duration
81d42a6f	my-cloud-build-trigger ⬈	master	78885ca ⬈	build-on-push-to-master	1/30/21, 12:37 AM	57 sec

Figure 6.7 – Summary of the build history specific to the Cloud Build trigger

To view the details of the build, click on the specific build. The details will show reference to steps that include the execution of the Dockerfile and the creation of a container image that is pushed to Container Registry (refer to *Figure 6.8*):

```
Step #0: Status: Downloaded newer image for tiangolo/uvicorn-gunicorn-fastapi:python3.7
Step #0:  ---> e2f19ac0b4e3
Step #0: Step 2/4 : COPY ./app /app
Step #0:  ---> 15bad8d7803d
Step #0: Step 3/4 : EXPOSE 8080
Step #0:  ---> Running in d494533d6248
Step #0: Removing intermediate container d494533d6248
Step #0:  ---> 5eed921f34ec
Step #0: Step 4/4 : CMD ["uvicorn", "main:app", "--host", "0.0.0.0", "--port", "8080"]
Step #0:  ---> Running in 82b4ee71960d
Step #0: Removing intermediate container 82b4ee71960d
Step #0:  ---> 3989bd42e41f
Step #0: Successfully built 3989bd42e41f
Step #0: Successfully tagged gcr.io/gcp-devops-2021/cloud-build-trigger:latest
```

Figure 6.8 – Log to build a container image and push to Container Registry

The newly created container can be found under **Container Registry** (refer to *Figure 6.9*):

Figure 6.9 – Viewing the container image in Container Registry

The end of the build log will show the deployment of the container image to Cloud Run. This will also include the newly created service URL to access the application (refer to *Figure 6.10*):

```
Step #2: Deploying container to Cloud Run service [cbt-cloud-run] in project [gcp-devops-2021] region [us-central1]
Step #2: Deploying...
Step #2: Setting IAM Policy............done
Step #2: Creating Revision.................................................................................................done
Step #2: Routing traffic......................................................done
Step #2: Done.
Step #2: Service [cbt-cloud-run] revision [cbt-cloud-run-00003-now] has been deployed and is serving 100 percent of traffic.
Step #2: Service URL: https://cbt-cloud-run-jejlhr4t3q-uc.a.run.app
```

Figure 6.10 – Log to deploy the container to Cloud Run

Navigate to the highlighted service URL to view the deployed application in Cloud Run (refer to *Figure 6.11*):

Figure 6.11 – Container image deployed in Cloud Run

This completes the hands-on lab where we deployed an application automatically to Cloud Run whenever a developer made a code change and pushed the code to the master branch. This illustrates an automatic CI/CD process that is built using GCP's native constructs such as Cloud Build and Container Registry.

Summary

In this chapter, we discussed two key services that are central to building a CI/CD workflow in Google. These are Cloud Build and Container Registry. Cloud Build is critical to build application code and output container images as build artifacts. Container Registry manages these build artifacts using the concepts of artifact management. The chapter went into in-depth details with respect to each of the services' key constructs and concluded with a hands-on lab where users can automatically deploy code to Cloud Run when a code change is detected by a configured trigger.

Google strongly recommends deploying applications using containers specifically against GKE, which is a key container deployment option apart from App Engine flexible and Cloud Run. The key concepts of GKE will be discussed in the next three chapters, which include understanding the core features of native Kubernetes, learning about GKE-specific features, and topics specific to hardening a GKE cluster.

Points to remember

The following are some important points to remember:

- Cloud Build can import source code from Google Cloud Storage, CSR, GitHub, or Bitbucket.
- Cloud builders are container images that run the build process.

- Google-managed builders are pre-built images that can be used to execute one or more build steps.

- Community-contributed builders are open source builders but not pre-built images and only source code is made available.

- The build configuration is a configuration file that encapsulates the steps to perform build-related tasks, written in `yaml` or `json` format.

- Manual invocation and automatic builds using triggers are the two main options to invoke the build process via Cloud Build.

- Cloud Build related logs are stored in Cloud Storage and Cloud Logging.

- Cloud Build Editor provides full control of Cloud Build resources.

- Cloud Build creates a specific Cloud Build SA (with minimal permissions assigned) for a project when the Cloud Build API is enabled on a project.

- Two common strategies to increase build speed are building leaner containers and using cached Docker images.

- Kaniko cache is a feature of Cloud Build where intermediate container image layers are directly written to Google's Container Registry.

- Cloud Build provides an option to increase the speed of the build by using a higher CPU VM.

- Unwanted files during the Cloud Build process can be ignored using the `gcloudignore` file.

- Container Registry is GCP's private container image registry service, which supports Docker Image Manifest V2 and OCI image formats.

- If `gcr.io` is used, then the default location is considered as *United States*.

- Storage Admin provides the ability to push and pull images from the Cloud Storage bucket associated with Container Registry.

- The gcloud credential helper and standalone credential helper are possible authentication methods that third-party clients can use to authenticate with Container Registry.

- Container Analysis is a service that provides vulnerability scanning and metadata storage for software artifacts.

- The Container Analysis API enables metadata storage and the Container Scanning API enables vulnerability scanning.

Further reading

For more information on Cloud Build and Container Registry, read the following articles:

- **Cloud Build**: `https://cloud.google.com/cloud-build`
- **Container Registry**: `https://cloud.google.com/container-registry`

Practice test

Answer the following questions:

1. Select all possible options that Cloud Build can import source code from (multiple):

   a) GitHub and Bitbucket

   b) Google Cloud Storage

   c) CSR

   d) None of the above

2. Cloud Build requires a build configuration file. Select the option that represents this:

   a) `cloudbuild.json, cloudbuild.xml`

   b) `build.json, build.yaml`

   c) `cloudbuild.json, cloudbuild.yaml`

   d) `build.json, build.xml`

3. Select the command that will configure Cloud Build to store an image in Container Registry during the build process:

   a) The `push` command

   b) The `docker put` command

   c) The `put` command

   d) The `docker push` command

4. Which of the following options can be used to store container images?

   a) Container Analysis

   b) Cloud Build

   c) Container Registry

   d) CSR

5. Select the option that stores trusted metadata used later in the authorization process:

    a) Container Registry

    b) Container Analysis

    c) Container Scanning

    d) Container Artifactory

6. Select the option that represents an intermediate image that is created by executing each instruction in a Dockerfile:

    a) Docker image

    b) Dockerfile

    c) Docker layer

    d) Docker daemon

7. Select the option that allows you to run multiple applications on the same physical hardware:

    a) OS

    b) Virtualization

    c) Hypervisor

    d) All of the above

8. Select all options that are applicable to Cloud Build:

    a) Managed service

    b) Serverless

    c) Both (a) and (b)

    d) None of the above

9. Which of the following is not a valid option that a user can provide in a build step (select one):

    a) name

    b) args

    c) env

    d) uniqueid

10. The build configuration file can be configured to store Cloud Build logs. Select the appropriate option to store logs:

    a) Cloud Storage

    b) Cloud Logging

    c) Both (a) and (b)

    d) None of the above

Answers

1. (a) – (b) and (c).

2. (c) – `cloudbuild.json`, `cloudbuild.yaml`.

3. (d) – The `docker push` command.

4. (c) – Container Registry.

5. (b) – Container Analysis.

6. (c) – Docker layer.

7. (b) - Virtualization.

8. (c) – Managed service and Serverless. Every serverless service is a managed service.

9. (d) – `uniqueid`. The right option is `id`.

10. (c) – Cloud Storage and Cloud Logging.

7

Understanding Kubernetes Essentials to Deploy Containerized Applications

The last two chapters (*Chapter 5*, *Managing Source Code Using Cloud Source Repositories*, and *Chapter 6*, *Building Code Using Cloud Build, and Pushing to Container Registry*) focused on Google Cloud services to manage source code via cloud source repositories, build code via Cloud Build, and create image artifacts using Container Registry. Given that the focus of this book is to deploy containerized applications, the next three chapters (from *Chapter 7*, *Understanding Kubernetes Essentials to Deploy Containerized Applications*, through to *Chapter 9*, *Securing the Cluster Using GKE Security Constructs*) are centered around essential concepts related to deploying containerized applications through Kubernetes, easy cluster management through **Google Kubernetes Engine** (**GKE**), and a rundown of key security features in GKE that are essential for hardening the Kubernetes cluster.

Kubernetes, or K8s, is an open source container orchestration system that can run containerized applications. Kubernetes originated as an internal cluster management tool from Google that it donated to **Cloud Native Computing Foundation** (**CNCF**) as an open source project in 2014. This specific chapter will focus on Kubernetes essentials that are required for containerized deployments. This includes understanding the cluster anatomy, and getting acquainted with Kubernetes objects, specifically related to workloads, deployment strategies, and constraints around scheduling applications. Google open sourced Kubernetes and donated it to CNCF. This specific chapter doesn't deep dive into setting up a Kubernetes cluster. It takes a significant effort and manual intervention to run the open source version of Kubernetes.

Google offers a managed version of Kubernetes called Google Kubernetes Engine, otherwise known as GKE. Essentially, the fundamentals of Kubernetes apply to GKE as well. However, GKE makes it easy to set up a Kubernetes cluster and includes additional capabilities that facilitate cluster management. The next chapter focuses on GKE core features and includes steps to create a cluster. However, this chapter essentially focuses and elaborates on the fundamentals of Kubernetes, which is also the core of GKE and helps to make the transition to GKE much easier.

This chapter introduces Kubernetes as the container orchestration of choice and provides details on the following topics:

- **Kubernetes**: A quick introduction
- **Kubernetes Cluster Anatomy**: Deep dives into the constructs that form a Kubernetes cluster along with the components that form the master control plane
- **Kubernetes Objects**: Provides a high-level overview of critical Kubernetes objects used to deploy workloads
- **Scheduling and interacting with Pods**: Details the constraints evaluated and interactions involved while scheduling applications on Kubernetes
- **Kubernetes deployment strategies**: Details potential deployment strategies, from the option that essentially recreates applications, via deployment options that ensure zero downtime, to the option that enables the shifting of a specific amount of traffic to the new application

Technical requirements

There are four main technical requirements:

- A valid **Google Cloud Platform** (**GCP**) account to go hands-on with GCP services: `https://cloud.google.com/free`.
- Install Google Cloud SDK: `https://cloud.google.com/sdk/docs/quickstart`.

- Install Git: `https://git-scm.com/book/en/v2/Getting-Started-Installing-Git`.

- Install Docker: `https://docs.docker.com/get-docker/`.

Kubernetes – a quick introduction

A **container** is a unit of software that packages code and its dependencies, such as libraries and configuration files. When compared to running applications on physical or virtual machines, a container enables applications to run faster and reliably across computing environments. Containers make it easier to build applications that use microservice design patterns. They are critical to the concept of continuous development, integration, and deployment as incremental changes can be made against a container image and can be quickly deployed to a compute environment of choice (that supports process isolation).

Given that containers are lean and easy to deploy, an organization might end up deploying its applications as several containers. This poses challenges as some of the applications might need to interact with one another. Additionally, the life cycle of the application should also be monitored and managed. For example, if an application goes down due to resource constraints, then another instance of the application should be made available. Similarly, if there is a sudden spike in traffic, the application should horizontally scale up and when traffic returns to normal, the application should subsequently scale down.

Scaling actions (up or down) should be provisioned automatically rather than manually. This creates a need for container orchestration and will be discussed as the next topic.

Container orchestration

Container orchestration is about managing the life cycle of containers, specifically in large dynamic environments. Container orchestration can control and automate tasks such as provisioning, deployment, maintaining redundancy, ensuring availability, and handling changing traffic by scaling up or down as needed.

In addition, container orchestration can also handle the following scenarios:

- Move containers from one host node to the other in case the host node dies.

- Set eviction policies if a container is consuming more resources than expected.

- Provision access to persistent storage volumes in case a container restarts.

- Secure interactions between containers by storing keys/secrets.

- Monitor the health of the containers.

Kubernetes traces its lineage from Borg – an internal Google project that is essentially a cluster manager that runs large-scale containerized workloads to support core Google services such as Google Search. Kubernetes was the next-generation cluster manager after Borg. The most popular concepts in Kubernetes came from Borg, such as Pods, services, labels, and ip-per-pod.

Alternative container orchestration options

Docker Swarm, Apache Mesos, OpenShift, and suchlike are a few alternatives for container orchestration outside Kubernetes. Docker Swarm is easy to get started and set up the cluster, but has limited features specific to scaling. Mesos is a cluster manager that is best suited to large systems and designed with maximum redundancy. It is complex in nature (in terms of features and configuration) and recommended for workloads such as Hadoop and Kafka, but is not suitable for mid- or small-scale systems.

The upcoming section summarizes the main features of Kubernetes.

Kubernetes features

The following are some of the key features in Kubernetes:

- **Declarative configuration**: Kubernetes administers the infrastructure declaratively, in other words, Kubernetes monitors the current state and takes the required action to ensure that the current state matches the desired state.

- **Automation**: Kubernetes' implementation of declarative configuration inherently supports automation. In addition, Kubernetes allows a wide range of user preferences and configurations. As a result, Kubernetes can automatically scale in and scale out containerized applications based on a myriad of conditions, with resource utilization or resource limits being a few of them.

- **Stateful and stateless**: Kubernetes supports both stateful and stateless applications. In the case of stateful applications, a user's state can be stored persistently. In addition, both batch jobs and daemon tasks are also supported.

- **Container management**: Kubernetes supports the features of infrastructure as service, such as logging, monitoring, and load balancing.

The next section outlines the structure of a Kubernetes deployment and deep dives into its key components and their capabilities.

Kubernetes cluster anatomy

A Kubernetes cluster is a collection of machines with compute power. These machines could be actual physical computers or could even be **virtual machines** (**VMs**).

In reference to cloud deployments, a Kubernetes cluster will be a collection of VMs. Each VM is termed a node. The nodes in a cluster are categorized as either *master* or *worker* nodes. Worker nodes run applications that are deployed in containers. The master node runs the control plane components that are responsible for coordinating tasks across the worker nodes.

Throughout this chapter, and for ease of reference, the node running the control plane components will be referred to as the master, and worker nodes will be referred to as nodes.

The master has the following responsibilities:

- It tracks information across all nodes in a cluster, such as applications or containers that a node is running.

- It schedules applications on nodes by identifying nodes based on requirements (such as resource, affinity, or Anti-Affinity constraints).

- It ensures that a desired number of instances are always running as per the deployment specifications and orchestrates all operations within the cluster.

Figure 7.1 shows an illustration of a Kubernetes cluster that includes both master and nodes. These are comprised of machines with compute power:

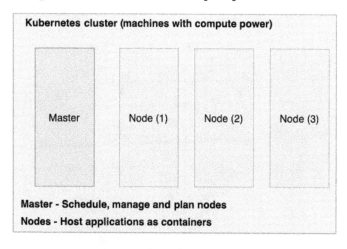

Figure 7.1 – Outline of a Kubernetes cluster

The master performs its responsibilities using a set of key components that form the **Kubernetes Control Plane** and will be detailed in the upcoming topic.

Master components – Kubernetes control plane

The Kubernetes control plane consists of components that make decisions with regard to operations within the cluster and respond to cluster-specific events. Events can include, but are not limited to, scaling up the number of instances with respect to the application if the average CPU utilization exceeds a specific configured threshold.

The following are the key components of the Kubernetes control plane:

- **kube-apiserver**: A frontend to Kubernetes for cluster interactions
- **etcd**: Distributed key-value stores for cluster-specific information
- **kube-scheduler**: Responsible for distributing workloads across nodes
- **kube-controller-manager**: Tracks whether a node or application is down
- **cloud-controller-manager**: Embeds cloud-specific control logic

Figure 7.2 shows an illustration of the components that run on the master and form the Kubernetes control plane:

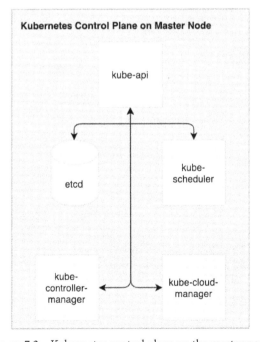

Figure 7.2 – Kubernetes control plane on the master node

The control plane components can be run on any machine in the cluster, but it is recommended to run on the same machine and avoid any user-specific containers. It's also possible to have multiple control planes when building a highly available cluster. Each of the key components is introduced in the upcoming sub-section.

kube-apiserver

kube-apiserver is the component of the control plane that exposes the Kubernetes API. This is the only component in the control plane that an end user or an external system/ Service can interact with. This component exposes the Kubernetes API through multiple pathways (such as HTTP, gRPC, and kubectl). All other components of the control plane can be viewed as clients to kube-apiserver. Additionally, kube-apiserver is also responsible for authentication, authorization, and managing admission control.

> **Authentication, authorization, and admission control**
>
> From a cluster standpoint, authentication is about who can interact with the cluster (this could be a user or service account); authorization is about what specific operations are permitted and admission control represents a set of plugins that could limit requests to create, delete, modify, or connect to a proxy. ResourceQuota is an example of an admission controller where a namespace can be restricted to only use up to a certain capacity of memory and CPU.

Any query or change to the cluster is handled by kube-apiserver. It is also designed to horizontally scale by deploying instances to handle incoming requests to the cluster.

etcd

etcd is the database for a Kubernetes cluster. etcd is a distributed key-value store used by Kubernetes to store information that is required to manage the cluster, such as cluster configuration data. This includes nodes, Pods, configs, secrets, accounts, roles, and bindings. When a get request is made to the cluster, the information is retrieved from etcd. Any create, update, or delete request made to the cluster is complete only if the change is reflected in etcd.

kube-scheduler

kube-scheduler is responsible for scheduling applications (encapsulated in Pod objects) or jobs onto nodes. It chooses a suitable node where the application can be deployed (but doesn't launch the application). To schedule an application, `kube-scheduler` considers multiple factors, such as the resource requirements of an application, node availability, affinity, and Anti-Affinity specifications. Affinity and Anti-Affinity specifications represent policy definitions that allow certain applications to be deployed against specific nodes or prevent deployment against specific nodes.

kube-controller-manager

kube-controller-manager monitors the state of the cluster through `kube-apiserver` and ensures that the current state of the cluster matches the desired state. `kube-controller-manager` is responsible for the actual running of the cluster, and accomplishes this by using several controller functions. As an example, a node controller monitors and responds when a node is offline. Other examples include the replication controller, namespace controller, and endpoints controller.

cloud-controller-manager

cloud-controller-manager includes controller functions that allow Kubernetes to be integrated with services from a cloud provider. The controller functions are responsible for handling constructs such as networking, load balancers, and storage volumes that are specific to the cloud provider.

The master receives a request to perform a specific operation and the components in the control plane schedule, plan, and manage the operations to be performed on the nodes. Kubernetes doesn't natively consist of out-of-the-box integration (say with Google or AWS). The operations on the nodes are carried out by a set of components that form the node control plane and will be detailed in the upcoming sub-section.

Node components

Nodes receive instructions from the master, specifically, the `kube-apiserver`. Nodes are responsible for running applications deployed in containers and establish communication between services across the cluster. The nodes perform these responsibilities by using a set of key components. These components are as follows:

- **kubelet**: An agent for Kubernetes that listens to instructions from `kube-apiserver` and runs containers as per the Pod specification provided

- **kube-proxy**: A network proxy that enables communication between services

- **container runtime**: Software that is responsible for running containers

Figure 7.3 shows an illustration of components that form the node control plane:

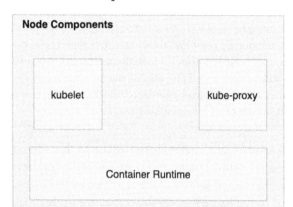

Figure 7.3 – Components of the node control plane

Node components run on each worker node in the cluster and provide the Kubernetes runtime environment. Each of the key components for a worker node is introduced in the upcoming topic.

kubelet

kubelet is an agent that runs on each node. When an action needs to be performed on a node, the kube-apiserver connects with the node through kubelet, the node's agent. kubelet listens for instructions and deploys or deletes containers when told to. kubelet doesn't manage containers that are not created by Kubernetes.

kube-proxy

kube-proxy is a network proxy that enables communication between services in the cluster based on network rules. The network rules allow communication with Pods from within or external to the cluster. kube-proxy runs on each node in the cluster.

container runtime engine

container runtime engine is the software that enables applications to be run in containers on the cluster. For example, the container runtime allows the DNS service and networking to run as containers. This includes the master plane components. Kubernetes supports multiple container runtimes, such as Docker, containerd, and CRI-O.

Kubernetes deprecating Docker as a container runtime engine

Based on the release notes for Kubernetes v1.20, `dockershim` will be deprecated and cannot be used from v1.22. `dockershim` is a module in `kubelet` and a temporary solution proposed by the Kubernetes community to use Docker as a container runtime. Due to the maintenance burden, `dockershim` will be deprecated and the Kubernetes community will only maintain the Kubernetes **Container Runtime Interface** (**CRI**). `containerd` and CRI-O are examples of a CRI-compliant runtime.

This completes a deep dive into Kubernetes cluster anatomy that specifically consists of components from the master control plane and components from the node control plane. Communication within the master control plane is driven by the `kube-api` server, which sends instructions to the `kubelet` on the respective nodes. `kubelet` executes the instructions sent by the `kube-api` server. *Figure 7.4* shows an illustration of the entire cluster anatomy:

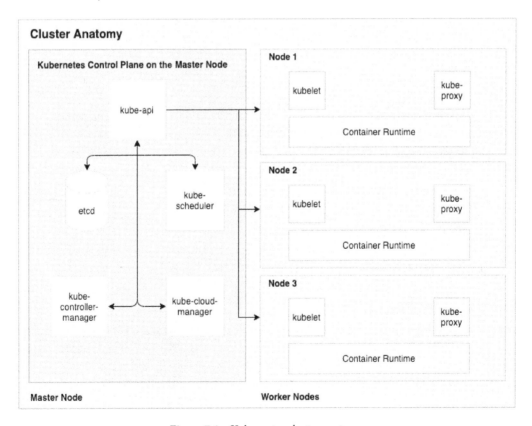

Figure 7.4 – Kubernetes cluster anatomy

It is important to understand that any interaction against a Kubernetes object, such as create, modify, or delete, can only be performed through the Kubernetes API. These operations on the object can also be performed through the CLI using the `kubectl` command. The next topic details the usage of the `kubectl` command.

Using kubectl

kubectl is a utility for controlling or executing operations in a Kubernetes cluster. `kubectl` is typically used by administrators. `kubectl` enables an action to be performed, such as get or delete, against a specific object type with a specific object name along with supported request parameters. `kubectl` communicates with `kube-apiserver` on the master and converts commands issued by the CLI into API calls. `kubectl` can be used to create Kubernetes objects, view existing objects, delete objects, and view/export configurations. The syntax structure of `kubectl` is as follows:

```
#kubectl syntax
kubectl [command] [type] [name] [flags]
#Example - Command to get specification of a specific pod
called 'my-pod' in yaml format
kubectl get pod my-pod -o=yaml
```

The first step involved in using the `kubectl` command is to configure the credentials of the cluster, such as the cluster name and its location. `kubectl` stores this configuration in a file called `config` and stores the file in a hidden folder called `.kube` in the home directory. The current configuration can be retrieved by using the `view` command:

```
# Get current config
kubectl config view
```

The actions on the cluster are executed using Kubernetes objects. Each object has a specific purpose and functionality. There are many such objects in Kubernetes. The upcoming section introduces the concept of Kubernetes objects and details the most frequently used objects.

Kubernetes objects

A Kubernetes object is a persistent entity and represents a record of intent. An object can be defined using the **YAML** configuration. It will have two main fields – spec and status. The object spec represents the specification, and the object state represents the desired state. Once the object is created, the Kubernetes system will ensure that the object exists as per the specified declarative configuration.

Kubernetes supports multiple object types. Each object type is meant for a specific purpose. The following are some critical Kubernetes objects that will be used throughout this chapter. This is not an exhaustive list:

- Pods – The smallest atomic unit in Kubernetes
- Deployment – Provides declarative updates for Pods and ReplicaSets
- StatefulSet – Manages stateful applications and guarantees ordering
- DaemonSet – Runs a copy of the Pod on each node
- Job – Creates one or more Pods and will continue to retry execution until a specified number of them terminate successfully
- CronJob – A job that occurs on a schedule represented by a cron expression
- Services – Exposes applications running one or more Pods

Deployment, ReplicaSet, StatefulSet, DaemonSet, Jobs, and CronJobs are specifically categorized as **Workload Resources**. All these workload resources run one or more Pods. This chapter details the abovementioned Kubernetes objects in the upcoming sub-sections. Please note that the information provided is not exhaustive from the aspect of an object's functionality, but provides an in-depth review of the object's purpose.

Pod

Pod is a Kubernetes object and is the smallest deployable compute unit in a Kubernetes cluster. Application code exists in container images. Container images are run using containers and containers run inside a Pod. A Pod resides inside a node.

A Pod can contain one or more containers. A Pod provides a specification on how to run the containers. The containers in a Pod share filesystem, namespace, and network resources. A Pod also has a set of ports or port ranges assigned. All containers in the Pod have the same IP address, but with different ports. The containers within the Pod can communicate by using the port number on localhost. The following is a declarative specification for a Pod that runs an `nginx` container:

```
apiVersion: v1
kind: Pod
metadata:
  name: my-nginx
spec:
```

```
containers:
- name: nginx
  image: nginx:1.14.2
  ports:
containerPort:80
```

The following is an equivalent imperative command to create a similar Pod:

```
kubectl run my-nginx –image=nginx
```

> **What is CrashLoopBackOff?**
>
> There are certain situations where a Pod attempts to start, crashes, starts again, and then crashes again; essentially, a condition reported by a Pod where a container in a Pod has failed to start after repeated attempts.

On the Kubernetes platform, Pods are the atomic units and run one or more containers. A Pod consists of multiple containers if they form a single cohesive unit of Service. The sidecar pattern is a common implementation of a Pod with multiple containers. These are popularly used in ETL-specific use cases. For example, logs of the `hello-world` container need to be analyzed in real time. `logs-analyzer` is a specialized application that is meant to analyze logs. If each of these containers is in their respective Pods as `pod-hello-world` and `pod-logs-analyzer`, the `logs-analyzer` container can get the logs of the `hello-world` container through a `GET` request. Refer to *Figure 7.5*:

Figure 7.5 – Communication between containers in different Pods

However, there will be minimal network latency since both the containers are in separate Pods. If both containers are part of the same Pod, `pod-hello-world-etl` forming a sidecar pattern, then the Pod will consist of two containers – `logs-analyzer` acting as the sidecar container that will analyze logs from another container, `hello-world`. Then, these containers can communicate on localhost because they are on the same network interface, providing real-time communication. Refer to *Figure 7.6*:

Figure 7.6 – Communication between containers in a sidecar pattern

Using a single Pod with multiple containers allows the application to run as a single unit and reduces network latency as the containers communicate on the same network interface. The following is a declarative specification of a Pod that runs multiple containers with the specific example as illustrated in *Figure 7.6*:

```
apiVersion: v1
kind: Pod
metadata:
  name: pod-hello-world-etl
spec:
  containers:
  - name: hello-world
    image: hello-world
    ports:
    - containerPort:8059
  - name: logs-analyzer
    image: custom-logs-analyzer:0.0.1
    ports:
    - containerPort:8058
```

Job and CronJob

Job and CronJob are workload resources. A job represents a task to execute a Pod. A job is completed if the task is executed successfully or, in other words, a Pod runs successfully to completion for a specified number of times. If a job is deleted, then Pods tied to the jobs are also deleted. If a job is suspended, then active Pods are deleted. Multiple jobs can be run in parallel. CronJob is a workload resource and is essentially a job that is set with a schedule through a cron expression.

Figure 7.7 brings together the examples related to a single container Pod, my-nginx, and the multiple container Pod, pod-hello-world-etl, and illustrates how these Pods can be potentially connected within a node:

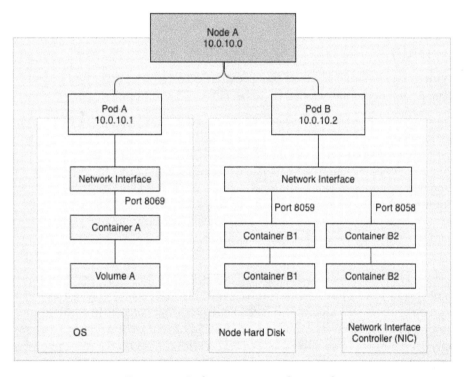

Figure 7.7 – Pod connectivity within a node

Pods are ephemeral in nature, and so is the storage associated with Pods. Hence, Pods are better suited for stateless applications. However, Pods can also be used for stateful applications, but in such cases, Pods should be attached to persistent storage or volumes. Pods are also meant to run a single instance of the application. Multiple instances of Pods should be used to scale horizontally. This is referred to as replication. So, Pods cannot scale by themselves.

> **Liveness, readiness, and start up probes**
>
> A liveness probe is used to check whether the application is running as expected and if not, the container is restarted. A readiness probe is used to check whether an application is up but also ready to accept traffic. A start up probe indicates when a container application has started. If a start up probe is configured, then this will disable the liveness and readiness checks until the start up probe succeeds. For more detailed information, refer to `https://kubernetes.io/docs/tasks/configure-pod-container/configure-liveness-readiness-startup-probes/`.

Kubernetes uses specific workload resources to create and manage multiple Pods. The most common ones are Deployment, StatefulSet, and DaemonSet, and these will be detailed in the upcoming sub-sections.

Deployment

A **Deployment** is a Kubernetes object that provides declarative updates for Pods and ReplicaSets. A Deployment is part of the Kubernetes API group called apps.

> **API groups**
>
> API groups are a way to extend the Kubernetes API. All supported API requests or future requests are placed in a specific group for easy categorization and this includes versioning. The most common group is the core group, also known as the legacy group. The core group is specified with `apiVersion` as `v1`. Pods fall under the core group. A Deployment falls under the apps group and is referred to with `apiVersion` as `apps/v1`.

A Deployment provides a declarative way to manage a set of Pods that are replicas. The deployment specification consists of a Pod template, Pod specification, and the desired number of Pod replicas. The cluster will have controllers that constantly monitor and work to maintain the desired state, and create, modify, or remove the replica Pods accordingly. Deployment controllers identify Pod replicas based on the matching label selector. The following is a declarative specification of a deployment that wraps three replicas of **nginx** Pods:

```
apiVersion: apps/v1
kind: Deployment
metadata:
  name: my-deploy
  labels:
```

```
      app: nginx
spec:
  replicas: 3
  selector:
    matchLabels:
      app:nginx
  template:
    metadata:
      labels:
        app: nginx
    spec:
      containers:
      - name: nginx
        image: nginx
        ports:
        - containerPort: 80
```

The following is a set of equivalent imperative commands to create a similar deployment:

```
kubectl create deployment my-deploy --image=nginx
kubectl scale -replicas=3 deployments/my-deploy
```

Deployments support autoscaling using the concept of **HorizontalPodAutoscaler (HPA)**, based on metrics such as CPU utilization. The following is the command to implement HPA. HPA will be discussed in detail as part of *Chapter 8, Understanding GKE Essentials to Deploy Containerized Applications*. This focuses on GKE:

```
kubectl autoscale deployment my-deploy --cpu-percent=80 --min=5
--max=10
```

A deployment can be updated using the rolling update strategy. For example, if the image version is updated, then a new ReplicaSet is created. A rolling update will ensure that the deployment will move the Pods from the old ReplicaSet to the new ReplicaSet in a phased-out manner to ensure 0% downtime. If an error occurs while performing a rolling update, the new ReplicaSet will never reach *Ready* status and the old ReplicaSet will not terminate, thereby enabling 0% downtime. Deployments and Pods are connected by labels. Each Pod is given a label. The deployment has a label selector. So, any updates to the deployments are rolled out to the Pods with matching labels.

A Deployment is well-suited for a stateless application, where a request will be served in a similar manner by either of the replica Pods. However, there is another Deployment resource that is stateful in nature and is called StatefulSets. This will be covered as the next topic.

StatefulSets

StatefulSets are a workload resource and a Kubernetes workload API object that is used to manage stateful applications, specifically, when an application requires its own unique network identifier and stable persistent storage. StatefulSets assign a unique identifier to Pods. StatefulSets can scale a set of Pods, but each replica is unique and has its own state. This means that each replica also has its own persistent volume. If the name of the StatefulSet is `sample`, then the Pod's name will be `sample-0`. If there are three replicas, then additional Pods called `sample-1` and `sample-2` will be created. This is completely different from deployment, where all Pods share the same volume.

The following is a declarative specification of a StatefulSet with three replicas:

```
apiVersion: apps/v1
kind: StatefulSet
metadata:
  name: sample
spec:
  selector:
    matchLabels:
      app: nginx
  serviceName: nginx
  replicas: 3
  updateStrategy:
    type: RollingUpdate
  template:
    metadata:
      labels:
        app: nginx
    spec:
      containers:
      - name: nginx
        image: ...
```

```
      ports:
      - containerPort: 80
      volumeMounts:
      - name: nginx-stateful-volume
        mountPath: ...
  volumeClaimTemplates:
  - metadata:
      name: nginx-stateful-volume
      annotations:

        ...

    spec:
      accessModes: [ "ReadWriteOnce" ]
      resources:
        requests:
          storage: 1Gi
```

If the StatefulSet is scaled down, then the last Pod in the StatefulSet will be removed (in other words, in reverse order). In the preceding example, if the replica count is reduced to two from three, then the `sample-2` Pod will be deleted. The StatefulSet supports rolling updates if there is any change. An old version of the Pod for a specific replica will be replaced when the new version of the Pod on that specific replica is back up. For example, `sample-0` will be replaced with a new version of `sample-0`. The next topic provides an overview of DaemonSets.

DaemonSets

DaemonSets is a workload resource that ensures that a copy of the Pod runs on every node or a certain subset of nodes in the cluster. Essentially, the controller for the DaemonSet creates a Pod when a node is created and deletes the Pod when the node is deleted. `kube-proxy` is a DaemonSet because a copy of it runs on each node in the cluster as part of the node control plane. Additional examples include the following:

- Running a log collector daemon on every node or a certain subset of nodes
- Running a cluster storage daemon on every node or a certain subset of nodes
- Running a node monitoring daemon on every node or a certain subset of nodes

To elaborate further, in the case of a log collection daemon, logs are exported from each node using a log collector such as `fluentd`. This can be done by a `fluentd` Pod and should be run on every node in the cluster. The following is a declarative specification of a log collection DaemonSet:

```
apiVersion: apps/v1
kind: DaemonSet
metadata:
  name: log-collector-daemon
  labels:
    app: daemon
spec:
  selector:
    matchLabels:
      app:log-collector
  template:
    metadata:
      labels:
        app: log-collector
    spec:
      containers:
      - name: fluentd
        image: quay.io/fluent/fluentd-kubernetes-daemonset
        ports:
        - containerPort: 9200
```

Like Deployments, the DaemonSet also supports rolling updates. So, if the DaemonSet is updated, then a new Pod is created and when the new Pod is up, the current Pod will be deleted.

The next topic discusses a Kubernetes object called Service. This is essential for establishing communication with an application from within the cluster and from outside the cluster.

Service

As previously mentioned, Pods are ephemeral in nature. Pods' IP addresses are not long-lived and can keep changing. This poses a challenge if an API request needs to be sent to the Pod's container using its IP address.

Kubernetes provides a stable abstraction point for a set of Pods called a **Service**. Every Service has a fixed IP address that doesn't change, and this gets registered with the cluster's built-in DNS. A Service identifies associated Pods using label selectors.

In addition, when a Service object is created, Kubernetes creates another object called **EndPoint**. The EndPoint object will maintain the list of all IPs for the Pods that match the label selector and is constantly updated as Pods are deleted and created. The Service object gets the current set of active Pods from the EndPoint object.

Figure 7.8 illustrates the interaction between the Service object, endpoint object, and the associated Pods based on the matching label selector:

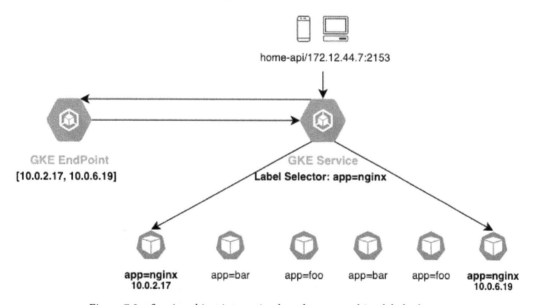

Figure 7.8 – Service object interaction based on a matching label selector

The following is a declarative specification that exposes a Service for a set of Pods. This allows the Pod to be accessed using the Service, since the Service is not ephemeral in nature and will have a fixed IP address:

```
apiVersion: v1
kind: Service
metadata:
  name: my-service
spec:
  selector:
    app: nginx
```

```
ports:
- protocol: TCP
  port: 80
  targetPort: 8080
```

The following is an equivalent imperative command that can expose Pods as a Service:

```
#Syntax
kubectl create <cluster-type> NAME [--tcp=port:targetPort]
kubectl create service clusterip nginx --tcp=80:80
```

There are four types of Service, and each Service type exposes Pods differently:

- ClusterIP
- NodePort
- LoadBalancer
- ExternalName

The preceding specification represents a Service of the ClusterIP type as that's the default Service type. This will be introduced as the next topic.

ClusterIP

ClusterIP is the default Service type. Each Service gets an IP that can only be accessed by other services within the cluster. This is essentially an internal IP, and hence the application inside the Pods cannot be accessed by public traffic or by an external Service that resides outside the cluster. The default Service type, if not specified, is ClusterIP. The preceding declarative specification is an example of a ClusterIP Service.

NodePort

NodePort is a Service type where the Service gets an internal IP that can be accessed by other services within the cluster. In addition, the NodePort Service gets a cluster-wide port. This port can be accessed by a Service that resides outside the cluster only if the request is sent to Node's IP address along with the cluster-wide port. Any traffic sent to the cluster-wide port will be redirected to the Pods associated with the Service. The following is a declarative specification that exposes a node port Service for a set of Pods:

```
apiVersion: v1
kind: Service
metadata:
```

```
  name: my-service
spec:
  type: NodePort
  selector:
    app: nginx
  ports:
  - protocol: TCP
    nodePort: 30200
    port: 80
    targetPort: 8080
```

LoadBalancer

LoadBalancer is a Service type where the Service gets an internal IP that can be accessed by other services within the cluster. In addition, the Service also gets an external IP address that allows the application to receive traffic from a Service that resides outside the cluster. This is facilitated by the public cloud load balancer attached to the Service. The following is a declarative specification that exposes a load balancer Service for a set of Pods:

```
apiVersion: v1
kind: Service
metadata:
  name: my-service
spec:
  type: LoadBalancer
  selector:
    app: nginx
  ports:
  - protocol: TCP
    port: 80
    targetPort: 8080
```

ExternalName

ExternalName is a Service type where the Service uses DNS names instead of label selectors. So, a request from an internal client goes to the internal DNS and then gets redirected to an external name. The following is a declarative specification for a Service of the `ExternalName` type:

```
apiVersion: v1
kind: Service
metadata:
  name: my-service
spec:
  type: ExternalName
  externalName: hello.com
```

Hence, a request from an internal client will go to `my-service.default.svc.cluster.local`, and then the request gets redirected to `hello.com`.

This completes an overview of the most common Service types in Kubernetes. One of the factors to consider while using services is to map services to Pods, otherwise known as Service resolution. Kubernetes has an add-on feature called **kube-dns**. `kube-dns` is a DNS server that is essentially a directory mapping of IP addresses against easy-to-remember names along with a record type:

The `kube-dns` server watches the API server for the creation of a new Service. When a new server is created, the `kube-dns` server creates a set of DNS records. Kubernetes is configured to use the `kube-dns` server's IP to resolve DNS names for Pods. Pods can resolve their Service IP by querying the `kube-dns` server using the Service name, the Pod's namespace, and the default cluster domain:

- If the Pod and Service are on the same namespace, then the Pod can resolve the Service IP by querying the `kube-dns` server using the Service name directly.
- If the Pod and Service are not on the same namespace, then the Pod can resolve the Service IP by querying the `kube-dns` server using the Service and the Service namespace.
- A Pod in any other namespace can resolve the IP address of the Service by using the fully qualified domain name, `foo.bar.svc.cluster.local`.

`kube-dns` maintains the following types of DNS record for Pods and services:

- Every Service defined in the cluster is assigned a DNS A record.
- Every named Pod in the cluster is assigned a DNS SRV record.

The following table represents `kube-dns` records where the hostname is `foo` and the namespace is `bar`:

Record Type	Hostname	Namespace	FQDN (Fully Qualified Domain Name)	IP Address/Port
A	foo	bar	foo.bar.svc.cluster.local	10.45.0.15
SRV	foo	bar	_http._tcp.foo.bar.svc.cluster.local	80

This concludes a high-level overview of specific Kubernetes objects, and this should provide a good basis for discussing GKE in the next chapter. There are several other objects, such as job, CronJob, volumes, and persistent volumes, but a deep dive on those will be beyond the scope of the book.

The next topic details several concepts with respect to scheduling and interacting with Pods.

Scheduling and interacting with Pods

A Pod is the smallest unit of deployment in a Kubernetes cluster that runs containerized applications. The `kube-scheduler` master control plane component is responsible for finding a suitable node for the Pod and includes interactions with other components of the control plane. In addition, `kube-scheduler` needs to consider multiple configuration options, such as NodeSelector, NodeAffinity, and PodAffinity, to find the right node for the Pod. This section details the interactions that happen during a Pod creation and details the factors that need to be considered while scheduling Pods.

Summarizing master plane interactions on Pod creation

A Pod is a workload that needs to be deployed in a Kubernetes cluster. A Pod needs to run on a node and will host an application. A Pod can be in various phases. The following is a summary of valid Pod phases:

- **Pending**: A Pod is accepted by the Kubernetes cluster, but is waiting to be scheduled.

- **Running**: A Pod is tied to a node and the container in the Pod is running.

- **Succeeded** or **Completed**: All containers in a Pod have terminated successfully and will not be restarted.

- **Failed**: All containers in the Pod have terminated and at least one container exited with a non-zero status or failure.

- **Unknown**: The state of the Pod cannot be obtained due to a communication error between the node where the Pod should be running.

Right from the time a request is received to create a Pod to the time the Pod is created, there is a series of interactions between the components of the master plane that will create the Pod on the worker node. The sequence of interactions is listed as follows. This reflects a scenario where a Pod is being created. The sequence of steps for other interactions, such as list or delete, or even other workloads, such as job or deployment, follow the same pattern:

1. `kube-apiserver` receives a request to create a Pod. The request can come from a `kubectl` command or a direct API interaction.

2. `kube-apiserver` authenticates and authorizes the incoming request.

3. Upon successful validation, `kube-apiserver` creates a Pod object but will not assign the newly created Pod object to any node.

4. `kube-apiserver` will update the information about the newly created Pod object against the `etcd` database and sends a response to the original request for Pod creation that a Pod has been created.

5. `kube-scheduler` continuously monitors and realizes that there is a new Pod object but with no node assigned.

6. `kube-controller` identifies the right node to put the Pod and communicates this back to `kube-apiserver`.

7. `kube-apiserver` updates the node for the Pod object against the `etcd` database.

8. `kube-apiserver` passes instructions to `kubelet` on the node (worker) to physically create the Pod object.

9. `kubelet` creates the Pod on the node and instructs the container runtime engine to deploy the application image.

10. `kubelet` updates the status back to `kube-apiserver` and `kube-apiserver` updates the `etcd` database.

This summarizes the interactions between master plane components when a request is sent to the Kubernetes cluster through `kubectl` or the Kubernetes client. The next sub-section focuses on critical factors that should be considered while scheduling Pods against the node.

Critical factors to consider while scheduling Pods

There are multiple factors that kube-scheduler considers when scheduling a Pod against a node. One such common factor is resource requests and maximum limits. A Pod optionally allows the specification of CPU/memory requests and sets the respective maximum limits on a container basis. These requests and limits at container level are summed up for a Pod and are used by kube-scheduler to determine the appropriate node for the Pod. kube-scheduler schedules a Pod on the node where the Pod's requests and limits are within the node's available capacity.

A Pod provides additional properties that exercise more control in forcing kube-scheduler to schedule Pods only if certain conditions are met. A node also provides properties that are considered during scheduling. The following are such properties:

- **NodeSelector**: Schedules a Pod against the node with matching label values
- **NodeAffinity**: Schedules a Pod against the node with matching flexible conditions; also considers Anti-Affinity conditions to avoid scheduling a Pod against specific node(s)
- **Inter-pod affinity and Anti-Affinity**: Schedules a Pod on nodes with Pods having matching attributes; also considers Anti-Affinity conditions that avoid scheduling Pods against specific node(s) that have Pods with specific attributes
- **NodeName**: Schedules a Pod against a very specific node
- **Taints and Tolerations**: Avoids scheduling Pods on nodes that are tainted, but can make an exception if tolerations are defined on the Pod

The upcoming sub-sections will detail the aforementioned attributes.

Node Selector

nodeSelector is a Pod attribute that forces kube-scheduler to schedule a Pod only against a node with a matching label and corresponding value for the label.

For example, consider a cluster where nodes in the cluster belong to different CPU platforms. The nodes are labeled with a label selector and an appropriate value indicating the CPU platform of the node. If there is a need to run a Pod on a node with a specific CPU platform, then the Pod attribute, nodeSelector, can be used. kube-scheduler will find a node that matches the nodeSelector specification on the Pod against the matching label on the node. If no such node is found, then the Pod will not be scheduled.

Figure 7.9 shows the use of `nodeSelector` in a Pod and its matching relevance to a node specification:

```
apiVersion: v1                          1    apiVersion: v1
kind: Pod                               2    kind: Node
metadata:                               3    metadata:
  name: my-pod                          4      name: node01
spec:                                   5      labels:
  containers:                           6        cpuPlatform: Skylake
  - name: nginx                         7
    image: nginx
  nodeSelector:
    cpuPlatform: Skylake
```

Figure 7.9 – Specifying a nodeSelector on a Pod that matches a node

In the preceding example, `kube-scheduler` will schedule the Pod on a node where the node label is `cpuPlatform` and the corresponding value is `Skylake`.

Node Affinity and Anti-Affinity

Node Affinity and Anti-Affinity are a set of preferences (very similar to the node selector) where a Pod can be scheduled against a node based on matching labels. However, the affinity and Anti-Affinity preferences are more flexible. A matching expression with a potential range of values can be specified, unlike `nodeSelector`, where only an exact value for a label match can be specified. These preferences are only considered during scheduling and are ignored during execution. This means that once a Pod is scheduled on a node, the Pod continues to run on the node even though the node labels have changed.

In addition, the Node Affinity and Anti-Affinity preferences can be set against two properties that serve as a hard or soft constraint. These two properties are as follows:

- **requiredDuringSchedulingIgnoredDuringExecution**: This is a hard limit where a Pod is scheduled only if the criterion is met.

- **preferredDuringSchedulingIgnoredDuringExecution**: This is a soft limit where the scheduler tries to deploy the Pod on the node that matches the specified criterion. The Pod is still deployed on a node even if a match is not found.

The following is a Pod specification involving the use of `nodeAffinity`. The Pod specification indicates that the Pod should not be scheduled on nodes with a specific CPU platform:

```
apiVersion: v1
kind: Pod
metadata:
  name: pod-node-anti-affinity
spec:
  containers:
  - name: my-pod
    image: nginx
  affinity:
    nodeAffinity:
      requiredDuringSchedulingIgnoredDuringExecution:
        nodeSelectorTerms:
        - matchExpressions:
          - key: cpuPlatform
            operator: Not In
            values:
            - Skylake
            - Broadwell
```

In the preceding example, `kube-scheduler` will not schedule the Pod on nodes where the CPU platform is either `Skylake` or `Broadwell`.

Inter-pod affinity and Anti-Affinity

This is an extension of Node Affinity with the same fundamentals. This specification allows the scheduling of Pods to nodes based on the labels on the Pods that are already running on the nodes. Similarly, Anti-Affinity will ensure that a Pod is not scheduled on a node if there are other Pods of specific labels running on a node.

The rules for Pod affinity and Anti-Affinity can be illustrated as follows:

- **pod-affinity**: Pod P should be scheduled on Node N only if Node N has other Pods running with matching rule A.
- **pod-anti-affinity**: Pod P should not be scheduled on Node N if Node N has other Pods running with matching rule B.

Figure 7.10 shows a Pod specification with Pod affinity and Anti-Affinity definitions:

```
apiVersion: v1
kind: Pod
metadata:
  name: with-inter-pod-affinity-anti-affinity
spec:
  affinity:
    podAffinity:
      requiredDuringSchedulingIgnoredDuringExecution:
      - labelSelector:
          matchExpressions:
          - key: app
            operator: In
            values:
            - webserver
            - elasticserver
    podAntiAffinity:
      preferredDuringSchedulingIgnoredDuringExecution:
      - weight: 100
        podAffinityTerm:
          labelSelector:
            matchExpressions:
            - key: app
              operator: In
              values:
              - database
```

Figure 7.10 – Pod definition with inter-pod and anti-pod affinity

In the preceding example, `kube-scheduler` will schedule Pods on nodes where other Pods that are already running on the node have matching labels that reflect `app` as either `webserver` or `elasticserver`. On the other hand, `kube-scheduler` will not attempt to schedule Pods on nodes where other Pods that are already running on the nodes have matching labels that reflect `app` as a database. In short, this Pod specification tries to schedule Pods on nodes that don't run database applications.

Node name

nodeName is an attribute that can be specified in a Pod definition file and is also the simplest way to specify constraints regarding node selection. The biggest limitation of this kind of specification is that it is an all-or-nothing proposition.

For example, if the node is available for scheduling, then the Pod can be scheduled on that specific node. However, if the node is not available, then the Pod doesn't have a choice of any other node. A Pod cannot be scheduled until the node can take further workloads. In addition, nodes can be ephemeral, specifically, when the nodes are VMs. So, specifying a node name might not be a good design to start with, and hence this method is the least preferred. The following is a Pod specification with the `nodeName` attribute:

```
apiVersion: v1
kind: Pod
metadata:
  name: my-pod
spec:
  containers:
  - name: nginx
    image: nginx
  nodeName: node01
```

In the preceding example, `kube-scheduler` will attempt to schedule Pods on nodes where the node name is `node01`.

Taints and tolerations

Node Affinity and Pod affinity are properties of a Pod for finding a set of nodes. Taint is a property of a node that can repel one or more Pods. A node is tainted with a specific effect, based on a defined combination of a key, operator, and optionally, a value attribute. Possible indications that a node may be tainted are as follows:

- **NoSchedule**: Indicates that no more Pods can be scheduled on this node

- **NoExecute**: Indicates that no more Pods can run on this node and existing ones should be terminated

- **PreferNoSchedule**: Indicates a soft limit that no more Pods can be scheduled on this node

Tolerations is a feature that allows Pods to be scheduled on nodes with matching taints. So, tolerations are a way to counter the impact of a taint.

The following is the CLI command to taint a node:

```
# Taint a node
kubectl taint nodes node01 sky=blue:NoSchedule
```

The following is a Pod specification that defines toleration against the tainted node, which makes the Pod still eligible to be scheduled against the node:

```
kind: Pod
metadata:
  name: my-pod
spec:
  containers:
  - name: nginx
    image: nginx
  tolerations:
  - key: "sky"
    value: "blue"
    operator: "Equal"
    effect: "NoSchedule"
```

This is a two-part example for tainting a node:

- The CLI command taints node01 by specifying not to schedule Pods with a matching label key-value pair as sky=blue.

- However, the Pod specification defines a toleration for node01.

So, the Pod can be potentially scheduled on node01 by kube-scheduler. This completes the deep dive into critical factors that need to be considered while scheduling Pods on nodes.

In a Kubernetes deployment, application changes in terms of new features or bug fixes are reflected by deploying updated container images. There are several strategies to enforce a change in deployment or apply a new deployment. These will be discussed in the upcoming section.

Kubernetes deployment strategies

If a change is required to horizontally scale an application by increasing the number of replicas or if a change is required to the application by updating the container image, then a change is required to the deployment specification in Kubernetes. This will lead to automatic updates, either resulting in deploying additional Pods to scale horizontally or deploying a new Pod with the updated image and replace the current running Pod.

Changes to deployment can either happen by applying an updated deployment spec or by editing an existing deployment or specifically updating the image on the deployment. All of these can be done through the kubectl commands. However, the strategy used to perform the deployment makes an immense difference in terms of how end users of the application are impacted. There are four specific deployment strategies. Each of these strategies offers a different use case. These are mentioned as follows and will be illustrated in detail in the upcoming sub-sections:

- Recreate
- Rolling update
- Blue/Green
- Canary

The first deployment strategy that will be detailed will be the *Recreate* strategy.

Recreate strategy

The **Recreate** strategy is a basic strategy and is also the most straightforward compared to other strategies. Essentially, the current running Pods are all destroyed or brought down first and then the desired number of Pods are brought up against a new ReplicaSet.

The following is an example snippet that illustrates a Recreate update:

```
[...]
kind: deployment
spec:
  replicas: 4
  strategy:
    type: Recreate
[...]
```

Based on the preceding example snippet, Kubernetes will first bring down all four running Pods on the current ReplicaSet. Following that, Kubernetes will create a new ReplicaSet and will start four new Pods. Refer to *Figure 7.11*:

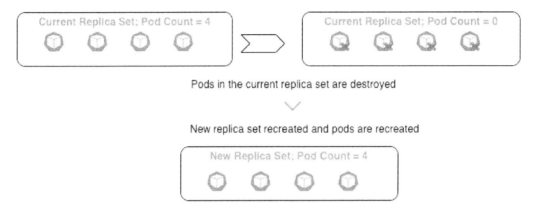

Figure 7.11 – Illustrating the 'Recreate' strategy in Kubernetes deployment

The Recreate strategy results in downtime as the application will remain unavailable for a brief period. This will result in disruptions and is therefore not a suggested strategy for applications that have a live user base. However, this strategy is used in scenarios where old and new versions of the application should or can never serve user traffic at the exact same time.

This completes the Recreate strategy. The clear downside is unavoidable downtime. This downside can be handled by another deployment strategy, called the rolling update strategy, and will be covered in the next sub-section.

Rolling update strategy

The **Rolling update** strategy enables the incremental deployment of applications with zero downtime. The current running Pod instances are gradually updated with a new Pod instance until all of them are replaced. The application stays available at all times. The rolling update strategy is the default deployment strategy in Kubernetes. However, this strategy doesn't exercise control in terms of specifying or having control over the amount of traffic directed to the new Pod instances versus old Pod instances.

The deployment also gets updated over time, and the process is time-consuming and gradual. There are specific fields that control the rolling update strategy, and these are detailed as follows, starting with **Max unavailable**.

Max unavailable

`.spec.strategy.rollingUpdate.maxUnavailable` is an optional field and refers to the maximum number of Pods that can be unavailable during the deployment process. This can be specified as an absolute number, or as a percentage of the desired Pods. If the field is not explicitly specified, then the default value is 25%. In addition, the default value is always considered if the value is explicitly specified as 0.

Let's consider an example. If the desired set of Pods is 5, and `maxUnavailable` is 2, this means that at any point in time, the total number of minimum Pods running across the old and new versions should be 3.

The next sub-section will cover **Max surge**. This indicates the maximum number of Pods that can exist at any time across current and new replica sets.

Max surge

`.spec.strategy.rollingUpdate.maxSurge` is an optional field and refers to the maximum number of Pods that can be created in addition to the desired number of Pods during the deployment process. This can be specified as an absolute number or a percentage of the desired Pods. If the field is not explicitly specified, then the default value is 25%. In addition, the default value is always considered if the value is explicitly specified as 0.

Let's consider an example. If the desired set of Pods is 5 and the maximum surge is 3, this means that the deployment process can get started by rolling out three new Pods and ensure that the total number of running Pods doesn't exceed 8 (desired Pods + maximum surge). If the maximum surge is specified in terms of a percentage and the value is set to 20%, then the total number of running Pods across old and new deployment will not exceed 6 (desired Pods + 10% of desired Pods).

The next sub-section will cover **Min Ready**. This indicates the minimum time that the container should run for, indicating the Pod to be ready.

Min ready

`.spec.minReadySeconds` is an optional field and refers to the minimum number of seconds that a newly created Pod should be in the ready state where containers are running without any failures or crashes. The default value, if not specified, is 0 and indicates that the Pod is ready as soon as it is created. However, if a value of 10 seconds is specified, for example, then the Pod needs to be in the ready state for 10 seconds without any containers failing in order to consider the Pod as available.

The next sub-section will cover **Progress Deadline**; the minimum wait time before concluding that a deployment is not progressing.

Progress deadline

`.spec.progressDeadlineSeconds` is an optional field and refers to the waiting period before a deployment reports that it has failed to progress. The default value, if not explicitly specified, is 600 (in seconds). If explicitly specified, then this value needs to be greater than `.spec.minReadySeconds`.

During a rolling update strategy, a new ReplicaSet is always created. New Pods are created in the new ReplicaSet and currently running Pods are gradually removed from the old ReplicaSet. The following is an example snippet that illustrates a rolling update strategy and includes the key fields that impact the way the rolling update will be performed internally:

```
[...]
kind: deployment
spec:
  replicas: 8
  minReadySeconds: 5
  strategy:
    type: RollingUpdate
    rollingUpdate:
      maxSurge: 4
      maxUnavailable: 50%
[...]
```

Based on the preceding example snippet, the following are the specific values that will be used to illustrate the example:

- Desired number of Pods = 8 Pods.

- Maximum surge = 4. At any point, the total number of running Pods across old and new running Pods cannot exceed 12.

- Maximum unavailable = 50% of desired = 4. At any point, there should be a minimum of 4 running Pods across old and new replica sets.

Kubernetes will create a new ReplicaSet and will launch 4 new Pods. Kubernetes will then wait for 5 seconds once the Pods have been created to consider whether the Pods are available. So, at this moment, the total number of Pods across old and new replica sets is 12, which is the maximum value allowed. This is illustrated in *Figure 7.12*:

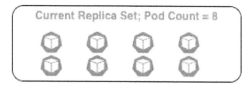 

Desired Number of Pods = 8

Figure 7.12 – Rolling update; creating Pods up to the maximum surge in a new ReplicaSet

Now, given that the minimum number of running Pods is four in this example, across the old and new replica sets, Kubernetes can potentially kill all eight of the old replica sets since it will still leave four in the new ReplicaSet. So, the core values are still not violated. This is illustrated in *Figure 7.13*:

Desired Number of Pods = 8

Figure 7.13 – Rolling update; removing Pods up to the maximum unavailable in the current ReplicaSet

Now, Kubernetes will launch four more new Pods in the new ReplicaSet and will reach the desired number of Pods as well. This is illustrated in *Figure 7.14*. This completes the rolling update, where the specified limits were met throughout the process:

Desired Number of Pods = 8

Figure 7.14 – Rolling update; creating new Pods up to the desired number in a new ReplicaSet

The rolling update strategy ensures zero downtime, but the downside is that there is no control in terms of the time taken for the deployment to complete, or no control in terms of the traffic going across old and new versions. The next strategy solves this specific downside.

Blue/Green strategy

In the case of the **Blue/Green** strategy, there will be two versions of the deployment running. That means that there are two replica sets, one ReplicaSet per deployment. However, each ReplicaSet will have a different set of labels that differentiate the Pods. Traffic to the Pods is sent through a Service. The Service will initially have labels that send traffic to the first deployment or ReplicaSet. The second deployment will also be running, but traffic will not be served. When the Service is patched, and the labels are updated on the Service, matching the labels of Pods on the second deployment, then traffic will be diverted to the second deployment without any downtime.

The following is an example of two running deployments in the Kubernetes cluster. In this example, the name of the deployment is demo-app. Both deployments are running the same application, but different versions of the application image. The difference in the deployment is also reflected by the Pod label selector, where the current version of the deployment has a label selector with the version as *blue*, whereas the new deployment has a label selector with the version as *green*.

The example is illustrated in *Figure 7.15*. The Service initially points to the deployment where the version is blue. This is because the label selector on the Service has the version as *blue* and matches the label selectors on the Pods in the current deployment. Hence, the incoming traffic to the Service is only handled by Pods in the *blue* deployment. The Pods in the *green* deployment are running, but they are not serving any incoming traffic:

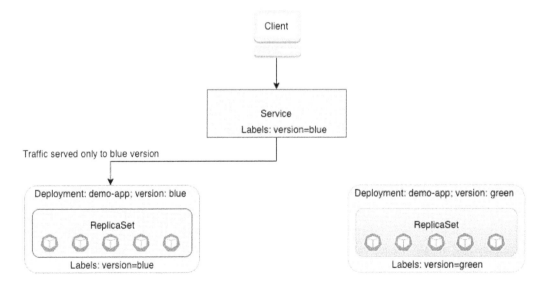

Figure 7.15 – Blue/Green deployment; traffic served only by the blue version

Figure 7.16 shows a snippet that reflects an update to the Service spec, where the Service label selector is updated to reflect the new version as green from blue:

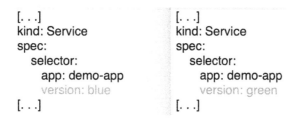

```
[. . .]                      [. . .]
kind: Service                kind: Service
spec:                        spec:
    selector:                    selector:
        app: demo-app                app: demo-app
        version: blue                version: green
[. . .]                      [. . .]
```

Figure 7.16 – Updating the Service specification to switch to a new deployment version

Figure 7.17 reflects how the traffic is served after the Service label selector is updated. In this case, incoming traffic will now be served by the Pods in the green deployment, as the Pod label selectors match those of the Service. The Pods in the blue deployment will no longer serve incoming traffic, although the Pods can continue to run:

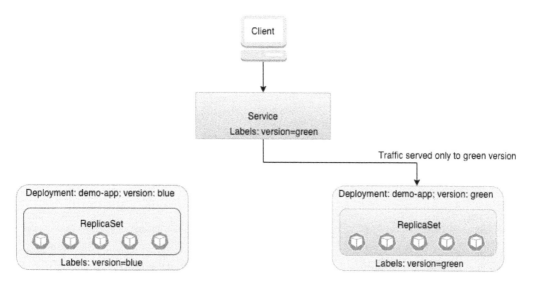

Figure 7.17 – Blue/Green deployment; traffic served only by the green version

Rolling out the deployment is as simple as updating the labels on the Service to point back to Pods matching the blue deployment.

Blue/Green deployment is alternatively known as red/black deployment, or A/B deployment. Although Blue/Green deployment provides control over the specific deployment against which traffic can be sent or rolled back, the downside is that double the number of applications are always running, thereby increasing infrastructure costs significantly. Also, it's an all-or-nothing scenario where a bug or issue in the application related to the updated deployment impacts all users of the application. This downside is solved by using the next deployment strategy, canary deployment.

Canary deployment

Canary deployment provides more control in terms of how much traffic can be sent to the new deployment. This ensures that a change in the application only impacts a subset of the users. If the change is not as desired, then it also impacts only a small percentage of total active users, thereby controlling customers' perceptions. Canary deployment is increasingly used in a continuous deployment process as it's a slow change where new features can be constantly added to active users, but in a controlled fashion.

Figure 7.18 illustrates a canary deployment where only 10% of the traffic is sent to the new deployment (version=green), whereas the remaining 90% is going to the current deployment (version=blue):

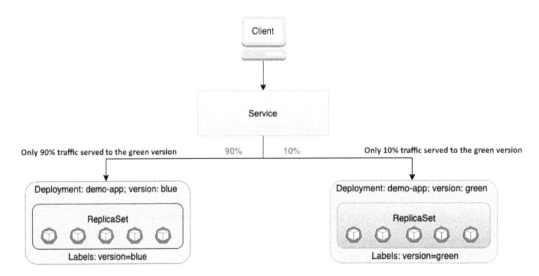

Figure 7.18 – Canary deployment; traffic sent to both versions based on the weighted percentage

Canary deployment is a true and reliable reflection of the continuous deployment model since a change to the application can flow through the CI/CD pipeline and can be deployed to production. In addition, deployment can also be targeted at just a specific set of users or a user base. This ensures that the new features are tested by live users (like beta users), but also ensures that a break in the new feature doesn't negatively impact the entire user base. Canary deployment is popularly implemented in the real world by using a resource such as Istio. Istio can split and route traffic between two versions based on predefined weights. As the new version becomes more stable, the traffic can gradually be shifted to the new deployment by changing the weighted percentage.

This completes a detailed illustration of the possible deployment strategies in Kubernetes. This also concludes a chapter that primarily focused on understanding the essential Kubernetes constructs for containerized deployments.

Summary

In this chapter, we discussed Kubernetes workloads in detail and considered Kubernetes as an option for deploying containerized applications. We learned about Kubernetes cluster anatomy, with a specific focus on understanding the key components that form the master control plane and the node control plane. In addition, we focused on learning key Kubernetes objects that are critical to deploying applications in the cluster, along with possible deployment strategies. Finally, we deep dived into how the master plane components interact while performing an action against an object such as Pod and discussed various factors involved in scheduling Pods onto Kubernetes nodes.

The next chapter focuses on the managed version of Kubernetes, called GKE, or GKE. The fundamental constructs of Kubernetes studied in this chapter, such as cluster anatomy or Kubernetes objects, are essentially the same for GKE. However, GKE makes cluster creation a lot easier and, in addition, GKE provides additional features for cluster management. Topics specifc to GKE, such as node pools, cluster configuration choices and autoscaling will also be detailed.

Points to remember

The following are some important points to remember:

- A node in a Kubernetes cluster is categorized as a *master* or *worker* node. The master node runs the control plane components.

- The key components of the Kubernetes control plane are `kube-apiserver`, `etcd`, `kube-scheduler`, `kube-controller-manager`, and `cloud-controller-manager`.

- It is recommended to run the control plane components on the same node and avoid any user-specific containers on that node.

- A highly available cluster can have multiple control planes.

- `kube-apiserver` handles any queries or changes to the cluster and can be horizontally scaled.

- `etcd` is a distributed key-value store used by Kubernetes to store cluster configuration data.

- `kube-scheduler` chooses a suitable node where an application can be deployed.

- `kube-controller-manager` runs several controller functions to ensure that the current state of the cluster matches the desired state.

- `cloud-controller-manager` includes controller functions that allow Kubernetes to integrate with services from a cloud provider.

- Key components of the worker node include `kubelet` (a Kubernetes agent that listens to instructions from `kube-apiserver` and runs containers as per the Pod specification), `kube-proxy` (a network proxy to enable communication between services), and container runtime (software responsible for running containers).

- Deployment, ReplicaSet, StatefulSet, DaemonSet, Jobs, and CronJobs are categorized as workload resources, and each run one or more Pods.

- A Pod is the smallest deployable unit in a Kubernetes cluster and can contain one or more containers that share filesystem, namespace, and network resources.

- Deployment provides a declarative way to manage a set of Pods that are replicas.

- StatefulSets manage stateful applications and can scale a set of Pods, but each replica is unique and has its own state.

- DaemonSets ensure that a copy of the Pod runs on every node or a certain subset of nodes in the cluster.

- The EndPoint object will maintain a list of all IPs for the Pods that match the label selector and is constantly updated as Pods are deleted and created.

- ExternalName is a Service type where the Service uses DNS names instead of label selectors.

- Critical factors to consider while scheduling Pods are NodeSelector, NodeAffinity, inter-pod affinity and Anti-Affinity, taints and tolerations, and NodeName.

- Possible Kubernetes deployment strategies are Recreate, Rolling update, Blue/Green, and Canary.

Further reading

For more information on GCP's approach to DevOps, read the following article:

- **Kubernetes**: https://kubernetes.io/docs/home/

Practice test

Answer the following questions:

1. A user changes the image of a container running in a Pod against a deployment in a Kubernetes cluster. A user updates the deployment specification. Select the option that describes the accurate behavior:

    a) The container image of the Pod tied to the deployment will get instantly updated and the running Pods will use the new container image.

    b) A new ReplicaSet will be created with the new image running inside a new Pod and will run in parallel with the older ReplicaSet with the older image.

    c) The current running Pod will stop instantly, and a new Pod will be created with the new image. There will be some downtime.

    d) A new ReplicaSet will be created with the new image running inside a new Pod and will gradually replace Pods from the old ReplicaSet.

2. Select the smallest unit of deployment in Kubernetes:

    a) Deployment

    b) Container

    c) Pod

    d) ReplicaSet

3. Select the object and the basis on which a Service object directs traffic:

    a) The Service object sends traffic to Deployments based on metadata.

    b) The Service object sends traffic to Pods based on label selectors.

    c) The Service object sends traffic to containers based on label selectors.

    d) The Service object sends traffic to Pods based on using the same name for the Pod and Service.

4. A Pod is in a ready state, but performing some actions internally when started, and is thereby unable to serve incoming traffic. Traffic from a Service is failing. Select the option that could be a potential solution:

    a) Configure a start up probe.

    b) Configure a liveness probe.

    c) Configure a readiness probe.

    d) None of the above.

5. There is a need to deploy multiple applications in a GKE cluster that could scale independently based on demand. Some of these applications are memory-intensive, some are I/O-intensive, and some are CPU-intensive. Select the option that represents the most appropriate cluster design:

    a) Select the majority category that applications fall under and create a cluster with either a CPU-intensive machine type, or memory-intensive or I/O-intensive.

    b) Create a cluster where the nodes have the maximum possible CPU and memory.

    c) Create a cluster with multiple node pools. Each node pool can be used to run a specific type of application with specific CPU, RAM, or I/O requirements.

    d) (b) and (c).

6. Which specific deployment option allows the testing of a new version of the application in production with a small percentage of actual traffic?

    a) Percentage deployment

    b) Rolling update

    c) Canary deployment

    d) Blue/Green deployment

7. Select the appropriate Service type where a Service gets an internal IP address:

    a) ClusterIP

    b) NodePort

    c) LoadBalancer

    d) All of the above

8. Which of the following deployment options enables running the last successful deployment on standby so that it could be used if the latest deployment has an issue? (Select all applicable options)

    a) Rolling update

    b) A/B deployment

    c) Canary deployment

    d) Red/black deployment

9. Which of the following controllers allows multiple development teams to use the same cluster, but with specific controls on the consumption of CPU and memory?

    a) Authorization controller

    b) `kube-controller-manager`

    c) ResourceQuota admission controller

    d) `cloud-controller-manager`

10. What is the function of a DaemonSet?

    a) It runs a specific Pod on every node in the cluster.

    b) It runs multiple copies of the specific Pod on every node in the cluster.

    c) It runs a specific Pod on every node in the cluster or a subset of selected nodes in the cluster.

    d) It runs multiple copies of the Pod on every node in the cluster or a subset of selected nodes in the cluster.

11. There is a specific requirement where Container C1 should be terminated if the memory or CPU currently utilized is three times more than the specified request limits. Select all possible options that match the specified requirements and should be added to the Pod spec:

    a) Requests: CPU=1000m, Memory=500Mi

       Limits: CPU=3000m, Memory=1250Mi

    b) Limits: CPU=3000m, Memory=1500Mi

       Requests: CPU=1000m, Memory=500Mi

    c) Requests: CPU=750m, Memory=1000Mi

       Limits: CPU=2250m, Memory=3000Mi

    d) Limits: CPU=1200m, Memory=500Mi

       Requests: CPU=3600m, Memory=1500Mi

12. A StatefulSet called `log-collector` consists of three replicas. Assume the Pods are labeled as `log-collector-0`, `log-collector-1`, and `log-collector-2`. The replica count is now scaled down to two replicas. Which of the following Pods will be deleted?

    a) The first Pod that was created in sequence will be deleted (`log-collector-0`).

    b) A random Pod will be deleted.

    c) The last Pod that was created will be deleted (`log-collector-2`).

    d) It's not possible to scale down a StatefulSet.

13. Select the option that depicts the reason for a CrashLoopBackOff error:

    a) Containers are terminated when an update is made to the Pod.

    b) A container in a Pod failed to start successfully following repeated attempts.

    c) Containers are terminated when an update is made to the deployment.

    d) None of the above.

14. Select the option that depicts the state where all containers in a Pod have terminated successfully and will not be restarted:

    a) Unknown

    b) Pending

    c) Completed

    d) Failed

15. Select the appropriate Service type where a Service gets a cluster-wide port:

    a) ClusterIP

    b) NodePort

    c) LoadBalancer

    d) All of the above

Answers

1. (d) – A new ReplicaSet will be created with the new image running inside a new Pod and will gradually replace Pods from the old ReplicaSet.

2. (c) – Pod.

3. (b) – The Service object sends traffic to Pods based on label selectors.

4. (c) – Configure a readiness probe.

5. (c) – Create a cluster with multiple node pools.

6. (c) – Canary deployment.

7. (d) – All of the above.

8. (b) and (d) – A/B and Red/Black is the same as Blue/Green.

9. (b) – ResourceQuota is an example of an admission controller where a namespace can be restricted to only use up to a certain capacity of memory and CPU. Each development team can work exclusively in its own namespace.

10. (c) – It runs a specific Pod on every node in the cluster or a subset of selected nodes in the cluster.

11. (b) and (c).

12. (c) – The last Pod that was created will be deleted (`log-collector-2`).

13. (b) – A container in a Pod failed to start successfully following repeated attempts.

14. (c) – Completed.

15. (b) – NodePort.

8
Understanding GKE Essentials to Deploy Containerized Applications

Kubernetes or K8s is an open source container orchestration system for automating the application deployment, scaling, and management of a cluster running containerized applications. The previous chapter introduced K8s fundamentals, including cluster anatomy, master plane components, Kubernetes objects (such as Pods and Services), workloads such as Deployments, StatefulSets, DaemonSets, and so on, and deep-dived into deployment strategies. However, setting up an open source Kubernetes cluster involves a lot of work at the infrastructure level and will also take a lot of time to set up. This also includes post-maintenance activities such as updating, upgrading, or repairing the cluster. GCP provides a compute offering that provides a managed Kubernetes or K8s environment called **Google Kubernetes Engine** (**GKE**).

The chapter introduces Google Kubernetes Engine as the managed Kubernetes option in GCP and uses the concepts introduced in *Chapter 7, Understanding Kubernetes Essentials to Deploy Containerized Applications*, to create a managed GKE cluster, deploy a containerized application into the cluster, and expose the application to be accessible from external clients. The chapter later details key GKE features, including the following topics:

- **Google Kubernetes Engine (GKE) – introduction**
- **GKE – core features**
- **GKE Autopilot – hands-on lab**

Technical requirements

There are four main technical requirements:

- A valid **Google Cloud Platform (GCP)** account to go hands-on with GCP services: `https://cloud.google.com/free`
- Install Google Cloud SDK: `https://cloud.google.com/sdk/docs/quickstart`
- Install Git: `https://git-scm.com/book/en/v2/Getting-Started-Installing-Git`
- Install Docker: `https://docs.docker.com/get-docker/`

Google Kubernetes Engine (GKE) – introduction

GKE is managed K8s and abstracts away the need to manage the master plane components from a user's standpoint. Creating a GKE cluster is much easier than creating a K8s cluster. This is because GKE cluster creation removes the need to manually create nodes, configure nodes and certificates, and establish network communication between the nodes. GKE also offers options to autoscale and manage auto-upgrades of the cluster's node software.

The following are the key features of a GKE cluster. These features differentiate GKE from open source Kubernetes or K8s:

- Fully managed and abstracts away the need for a user to provide underlying resources.
- Uses a container-optimized OS, an OS that is maintained by Google and is built to scale quickly with minimal resource requirements.

- Supports auto-upgrade and provides options to either get the latest available features or a more stable version without manual intervention.

- Provides the ability to auto-repair nodes by continuously monitoring the status of the nodes. If unhealthy, the nodes are gracefully drained and recreated.

- Automatically scales the cluster by adding more nodes as needed.

In addition to the preceding list, here are some additional key features that are available in K8s but need to be added and explicitly maintained as add-ons. These come as standard with GKE, thus making GKE a more viable and preferred option when compared to K8s:

- Load balancer – GKE provides a HTTP(S) load balancer.

- DNS – GKE implements service discovery and provides a managed DNS.

- Logging, monitoring, and dashboard – GKE provides these features built in due to its integration with Google Cloud operations.

Until recently, GKE offered only one mode of operation called **Standard** (also referred to as default). The Standard mode allows users to select the configurations needed to run workloads such as the node's machine type. This mode also allows you to select security configuration features, provides the ability to group nodes that run similar workloads, provides options to configure networking, and so on. Essentially, creating a cluster through GKE Standard mode is much easier than in open source K8s, but there is still a learning curve.

GKE recently introduced a new mode of operation called Autopilot. Autopilot has many of the configurations pre-selected and essentially creates a production-grade cluster that is hardened from a security standpoint. There are a few options to configure but, most importantly, the nodes are provisioned only when workloads are deployed. Autopilot mode will be discussed in detail later in this chapter through a hands-on lab.

> **Important note**
>
> The current chapter focuses on *Standard* mode unless explicitly specified. This will help you to understand the available options while creating a GKE cluster and provides insights into GKE features. Later in this chapter, the Autopilot mode will be elaborated on, calling out the key differences between the Standard and Autopilot modes, along with a hands-on lab.

GKE provides seamless integration with multiple service offerings from GCP. GKE provides options to automate deployment by building code stored in a source code repository using Cloud Build, which results in private container images that could be stored in Google's Container Registry. In addition, access to the cluster and the ability to configure GKE cluster options can be controlled via Google's **Identity and Access Management** (**IAM**). GKE integrates with GCP's network offerings as a GKE cluster is created as part of Google's Virtual Private Cloud or VPC. GCP provides insights into a GKE cluster and its resources as GKE integrates with Google's Cloud operations, a suite of tools from Google aimed at providing integrated services related to monitoring and logging.

We will start by creating a GKE cluster through a step-by-step process. This will provide an insight into the possible configuration options. Once the cluster is created, the user will be able to deploy an application through the concept of a Deployment and expose the application through the concept of a Service. The application runs inside a container wrapped by a Pod. The Deployment specification will manage the Pod. The Pod is then exposed using the concept of a Service. The concepts of Pods, Deployments, and services are K8s fundamentals that were discussed in *Chapter 7, Understanding Kubernetes Essentials to Deploy Containerized Applications*, and these concepts will be put into action on an actual GKE cluster.

Creating a GKE cluster

There are multiple ways to create a GKE cluster – the Cloud Console, CLI, or REST. To create a cluster, the user or the service account should have one of the following pre-defined roles: Kubernetes Engine Admin or Kubernetes Engine Cluster Admin.

The following is a step-by-step process to create a GKE cluster from the Google Cloud Console. The mode of operation will be **Standard** in this specific example:

1. Navigate to the GCP Console and select the compute service – **Kubernetes Engine**.

2. Select the option to create a cluster and choose the **Standard** mode.

3. Enter the name for the cluster as `my-first-cluster`.

4. Leave the default selections for the rest of the options. Refer to *Figure 8.1*:

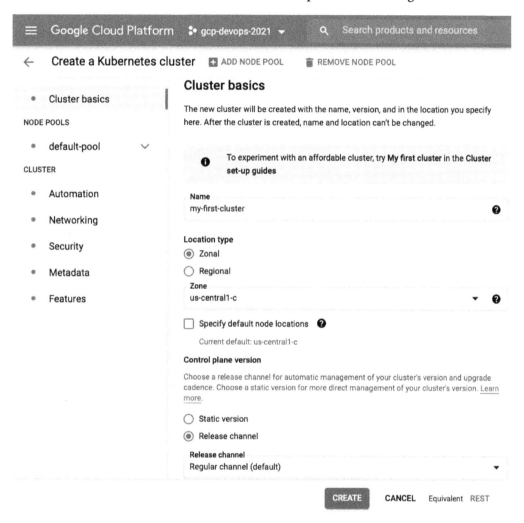

Figure 8.1 – Creating a GKE Cluster from the GCP Console

5. Select the option to **CREATE** the cluster. This will initiate the cluster creation process.

6. The newly created cluster will be displayed on the cluster home page. Refer to
    Figure 8.2:

Figure 8.2 – The GKE cluster list page displays the newly created cluster

The newly created cluster used the default options. Nothing really was changed during
the cluster creation except for the cluster name. The following are some important points
to know when a GKE cluster is created with default options. Each of the default options
mentioned in the following list can be explicitly changed during cluster creation:

* The default **Location type** of the cluster is **Zonal**. **Location type** refers to the cluster
  based on availability requirements. The options are **Zonal** and **Regional**.

* The default **Control plane zone** is us-central1-c. This indicates the zone where
  the control plane components are created.

* The default **Node Location** is us-central1-c. This indicates where the nodes are
  created. Multiple locations within a region can be selected to form a cluster where
  the location type is a multi-zonal cluster.

* The default **Control plane version** is **Release channel**. **Control plane version**
  provides options to signify the cluster version. The cluster version is indicative of
  the preferred feature set in terms of stability.

* The **default size** of the cluster is 3, indicating the number of worker nodes. The
  cluster, by default, only has 1 node pool. It's important to note that the cluster size
  doesn't include the master node count. Customers only pay for the worker nodes.
  The master node and the associated master plane components are entirely managed
  by GKE.

* The **default node pool** is named default-pool. A node pool is a collection
  of VMs.

- The default node pool consists of 3 nodes and the machine type for the node is `e2-medium` (2 vCPU, 4 GB memory).

- The default **Maintenance Window** is **anytime**. This implies that GKE maintenance can run at any time on the cluster. This is not the preferred option when running production workloads.

- The default cluster type based on networking is **Public cluster** and the default **VPC network** is **default**. This indicates how clients can reach the control plane and how applications in the cluster communicate with each other and with the control plane.

- Advanced networking options such as **VPC-native traffic routing** and **HTTP Load Balancing** are *enabled* by default. These options are discussed in detail in the sub-section *Networking in GKE*, later in this chapter

- **Maximum pods per node** defaults to `110`.

- The security feature **Shielded GKE Node** is *enabled*. This feature provides strong cryptographic identity for nodes joining a cluster and is discussed in detail as part of *Chapter 9, Securing the Cluster Using GKE Security Constructs*.

- Cloud operations for GKE are enabled and are set to **System, workload logging and monitoring**. This feature aggregates logs, events, and metrics for both infrastructure and application-level workloads.

A cluster can also be created from the **Command-Line Interface (CLI)**. The following is the CLI command to create a cluster with default options. The default options used in the CLI are the same as the default options used while creating a cluster from the console as described previously. One significant difference, however, is that it is mandatory to explicitly specify a zone while executing through the CLI. However, the zone is auto-filled in the UI unless modified:

```
# Create a GKE cluster with default options
gcloud container clusters create my-first-cli-cluster --zone
us-central1-c
```

This CLI command can be run from the terminal window of your local machine, which has Google Cloud SDK installed and configured. Alternatively, the CLI command can also be executed using Google Cloud Shell, activated through the Google Cloud Console.

Given that a GKE cluster is created, the next step is to deploy an application onto the GKE cluster and expose the application to an external client. This is discussed as the next topic.

GKE cluster – deploying and exposing an application

In *Chapter 6, Building code using Cloud Build, and Pushing to Container Registry*, we created a container image, and the container image was deployed using Cloud Run. In this chapter and in this sub-section, we will reuse this image and deploy it to the newly created GKE cluster by creating appropriate workloads. Once the application is deployed, the application will be exposed via a Service so that the application can be reached via an external client such as a web browser.

> **Important note**
>
> For continuity from an example standpoint, we will be using the container image created in *Chapter 6, Building code using Cloud Build, and Pushing to Container Registry* – `gcr.io/gcp-devops-2021/cloud-build-trigger`. It's recommended to use an appropriate container image of your choice that you have access to. For example, if you followed the step-by-step instructions in *Chapter 6, Building code using Cloud Build, and Pushing to Container Registry*, and ended up creating a container image in your project, you can reuse the same image in this chapter.

We will deploy the application and expose the application in two different ways:

- GKE Console
- The CLI approach via Cloud Shell

It's important to note that a cluster is typically deployed in most cases through the command line. However, we will first explore the GKE Console approach as this will give us insights into the available configuration options. This is covered as the next topic.

GKE Console

The first step is to deploy the application to the GKE cluster through the GKE Console.

Deploying an application to the GKE cluster

The following is the step-by-step process to deploy an application through the GKE Console:

1. Navigate to the **Clusters** page in the **Kubernetes Engine** section of the GCP Console.

2. Select the cluster that was previously created – `my-first-cluster`.

3. On the left-hand pane, select the section **Workloads**. From a GKE perspective, workloads refer to Deployments, StatefulSets, DaemonSets, Jobs, and CronJobs. There are no workloads at this moment and the current state will be as shown in *Figure 8.3*:

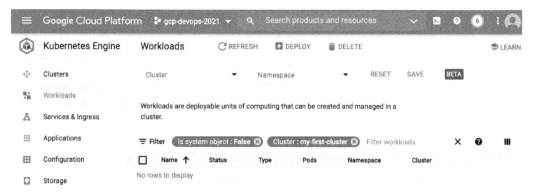

Figure 8.3 – The Workloads section of a newly created cluster

4. Create a workload by selecting the **DEPLOY** option. This action allows you to create a Deployment object in a two-step process.

5. The first step to create a Deployment is to define the containers required for the Deployment. Select the container image created in *Chapter 6*, *Building code using Cloud Build, and Pushing to Container Registry*. For this example, select the container image `gcr.io/gcp-devops-2021/cloud-build-trigger`. Refer to *Figure 8.4*. Optionally, add environment variables for the container and click on **Done**:

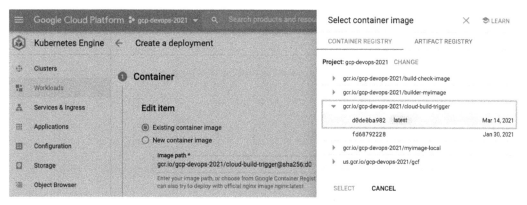

Figure 8.4 – Selecting container image while defining a container for Deployment

6. Optionally, multiple containers can be added to the Pod by using the **ADD CONTAINER** option. Refer to *Figure 8.5*:

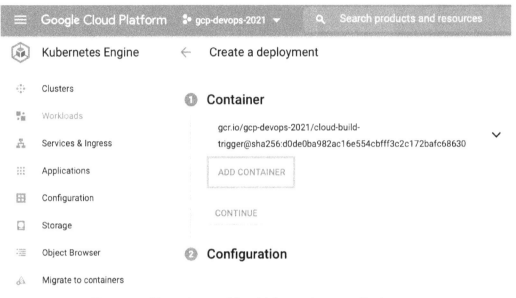

Figure 8.5 – The option to add multiple containers to a Deployment

7. The second step in creating a Deployment is to configure the Deployment. This includes specifying the application name, namespace, labels, and the cluster to which the application should be deployed. For this specific example, set **Application name** as `hello-world`, **Namespace** as `default`, **Labels** with **Key** as `app` and **Value** as `hello-world`, and select the cluster called `my-first-cluster`. Refer to *Figure 8.6*:

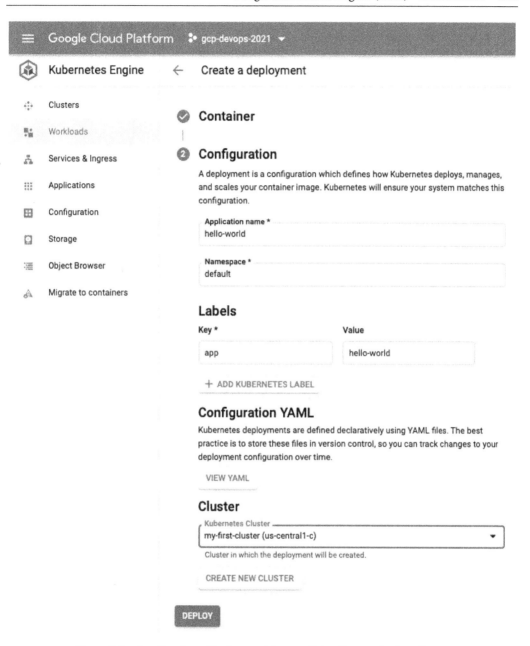

Figure 8.6 – Configuring a Deployment by specifying the required attributes

8. Before selecting the **DEPLOY** option, the configuration YAML can be viewed by selecting the **VIEW YAML** option as shown in *Figure 8.6*. By default, the number of replicas is defined as 3. This can optionally be changed to the desired replica count.

9. Initiate the deployment creation process by selecting the **DEPLOY** option.

10. The newly created Deployment – `hello-world` – will be displayed as follows. This Deployment created three replicas with the same image. Refer to *Figure 8.7*:

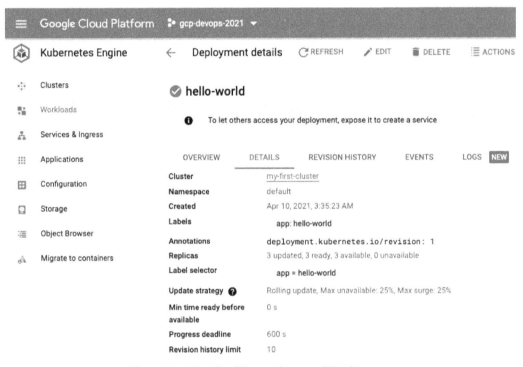

Figure 8.7 – Details of the newly created Deployment

It is important to note that the newly created Deployment – `hello-world` – cannot be accessed from external clients (such as a web browser or through a `ping` command) as the Deployment is not exposed as a Service. However, the application can still be tested by using the `port-forward` option. The CLI commands required to execute this option are shown in the following snippet. These commands can be executed through Google Cloud Shell:

```
# Connect to the cluster - 'my-first-cluster'
gcloud container clusters get-credentials my-first-cluster
--zone us-central1-c --project gcp-devops-2021
```

```
# Find the list of pods for the deployment hello-world
kubectl get pods


# For a specific pod, create a port-forward option to access
the application running inside the pod


kubectl port-forward hello-world-6755d97c-dlq7m 10080:8080
```

Once the preceding port-forward command is executed, traffic coming on 127.0.0.1:10080 will be forwarded to port 8080. Port 8080 is the container port related to the hello-world Deployment. Refer to *Figure 8.8*:

Figure 8.8 – Forwarding traffic to a container inside a Pod

To test whether traffic is getting forwarded, open another Cloud Shell window and run the curl command as shown. This will do a REST call invocation against the application running inside the container of a Pod. Refer to *Figure 8.9*:

Figure 8.9 – Result of accessing the application in a Pod through port-forwarding

Alternatively, you can also use the web preview option on port 10080 in Cloud Shell to view the application. Given that the application is now deployed and is working as expected, the next step is to expose the application as a Service.

Exposing the application as a Service

The following is a step-by-step process to expose the application as a Service through the GCP Console:

1. Navigate to the **Clusters** page in the **Kubernetes Engine** section of the GCP Console.

2. Select the cluster that was previously created – my-first-cluster.

3. Select the Deployment that was previously created – hello-world.

4. Under the **Actions** menu on the deployment details page, select the **EXPOSE** option. This will open a pop-up window where **Port**, **Target port**, **Protocol**, and **Service type** need to be selected.

5. Enter **Port** as 80 (this represents the port where the Service will be listening for incoming traffic), **Target port** as 8080 (this is the port the container will be listening on), **Protocol** as TCP, and **Service type** as Load balancer. Select the **EXPOSE** option. Refer to *Figure 8.10*:

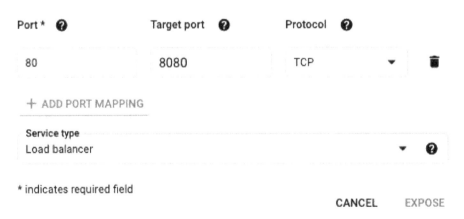

Figure 8.10 – Specifying port mapping to expose a Pod as a Service of type Load balancer

6. Once the Pod is exposed, a Service will be created as shown in the following screenshot. Given the Service is of type **LoadBalancer**, the Service will have an external endpoint. Refer to *Figure 8.11*:

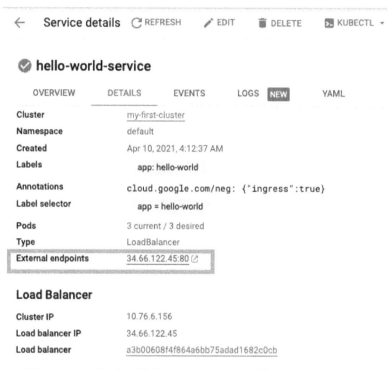

Figure 8.11 – The LoadBalancer Service created by exposing the Pod

7. Select the external endpoint. This will open the application in the browser as shown in the following screenshot. This essentially is the output of deploying the application to the GKE cluster. The output is the same as the output in *Chapter 6, Building code using Cloud Build, and Pushing to Container Registry*, when the same container image was deployed to Cloud Run. Refer to *Figure 8.12*:

```
{"Hello":"World"}
```

Figure 8.12 – Output of accessing the application through the load balancer Service

This completes the topic on deploying an application to the GKE cluster and exposing the application via a load balancer Service through the GKE Console. The next sub-section essentially works on a similar example but provides insights on how the same thing can be done through Cloud Shell using the CLI approach.

The CLI approach via Cloud Shell

In this sub-section, we will deploy an application and expose the application as a load balancer Service through the CLI using Cloud Shell. We will use the same cluster as was previously created – `my-first-cluster`. It is also recommended to use the container image created as part of the exercise in *Chapter 6*, *Building code using Cloud Build, and Pushing to Container Registry*. For this example, the container image `gcr.io/gcp-devops-2021/cloud-build-trigger` will be used.

Deploying an application to the GKE cluster

The following is the step-by-step process to deploy an application via Cloud Shell:

1. Open Cloud Shell and connect to the cluster using the following CLI command:

    ```
    # Connect to the cluster
    gcloud container clusters get-credentials my-first-
    cluster --zone us-central1-c --project gcp-devops-2021
    ```

2. Create a new file called `hello-world-cli.yaml` with contents as follows. This file essentially creates a Deployment that has the container and respective image to be deployed. The replica count is also specified and in this case, is 1:

    ```
    apiVersion: "apps/v1"
    kind: "Deployment"
    metadata:
      name: "hello-world-cli"
      namespace: "default"
      labels:
        app: "hello-world-cli"
    spec:
      replicas: 1
      selector:
        matchLabels:
          app: "hello-world-cli"
      template:
    ```

```
  metadata:
    labels:
      app: "hello-world-cli"
  spec:
    containers:
    - name: "cloud-build-trigger-sha256-1"
      image: "gcr.io/gcp-devops-2021/cloud-build-
trigger:latest"
```

3. Create the Deployment by running the following command:

```
kubectl apply -f hello-world-cli.yaml
```

Once the Deployment is created, the Deployment and its respective Pod can be queried as follows through the CLI. Please note that this Deployment will create only one Pod. Refer to *Figure 8.13*:

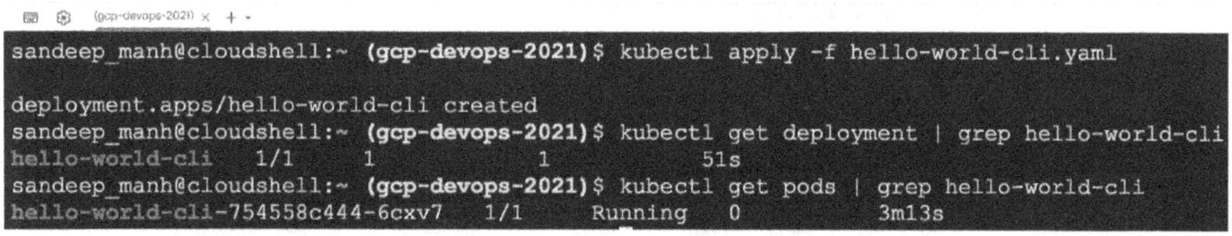

Figure 8.13 – Querying the Deployment through the CLI

The deployed application cannot be accessed through an external client. However, the port-forward approach explained in the previous sub-section can be exactly applied in this context as well. Given that the application is now deployed, the next step is to expose the application as a Service.

Exposing the application as a Service

The following is the step-by-step process to expose the application as a Service through Cloud Shell:

1. Create a new file called `hello-world-cli-service.yaml` with a definition as follows. This will create a load balancer Service that will expose a Pod with matching label selectors:

```
apiVersion: v1
kind: Service
metadata:
```

```
    name: hello-world-cli-service
spec:
  type: LoadBalancer
  selector:
    app: hello-world-cli
  ports:
  - protocol: TCP
    port: 80
    targetPort: 8080
```

2. Create the load balancer Service by running the following command:

    ```
    kubectl apply -f hello-world-cli-service.yaml
    ```

3. Once the Service is created, a load balancer will be created with an external
    endpoint. As per the Service definition, the Service will listen to traffic on port 80
    and will forward the traffic to the container on port 8080. The external endpoint of
    the Service can be found out by querying the Service as follows. Refer to *Figure 8.14*:

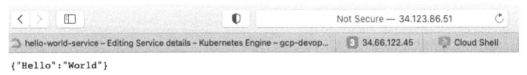

Figure 8.14 – Query the load balancer Service to fetch the external endpoint

4. Access the external endpoint through a browser window. The output will be the
    same as the output from *Chapter 6, Building code using Cloud Build, and Pushing to
    Container Registry*, or the output from the application deployed in GKE through the
    console. This is because we are using the same image. Refer to *Figure 8.15*:

| < > | ▢ | | ◑ | | | Not Secure — 34.123.86.51 | | ↻ |

| ⟳ hello-world-service – Editing Service details – Kubernetes Engine – gcp-devop... | | 🅱 34.66.122.45 | | Cloud Shell |

```
{"Hello":"World"}
```

Figure 8.15 – Viewing the output of the load balancer Service via an external endpoint

This concludes this section, which introduced GKE and took a deep dive into the step-by-step process to create a GKE cluster, deploy an application to the cluster, and expose the deployed application as a Service to be accessed by external clients. Essentially, the output of this approach is the same as the output from the console approach. The goal is to understand the process of creating a cluster, deploying workloads, and exposing the workloads through a Service via the CLI.

The concepts used while creating the cluster or deploying the application are the same concepts that form the fundamentals of K8s (learned about in *Chapter 7*, *Understanding Kubernetes Essentials to Deploy Containerized Applications*). However, the cluster creation is much simpler in nature since the maintenance of the master plane components is completely abstracted and is not the responsibility of the user. The upcoming section focuses on core GKE features and possible cluster types, and provides an introduction to integration with networking and cloud operations in GKE.

GKE – core features

This section covers the following topics. These topics will provide a considerable amount of information, which is required to build a good understanding and working knowledge of GKE. Most of these GKE concepts are an extension of topics learned about in the Kubernetes section. The topics that will be covered are as follows:

- GKE node pools
- GKE cluster types
- Autoscaling in GKE
- Networking in GKE
- Cloud operations for GKE

The first of the GKE constructs that will be detailed in the upcoming sub-section is GKE node pools.

GKE node pools

Nodes (that is, worker nodes) in a Kubernetes cluster deploy workloads. The nature of workloads deployed across all nodes might not be the same. Some workloads might be CPU-intensive, others might be memory-intensive, and some might need a minimum version of the CPU platform. Workloads can also be fault-tolerant batch jobs or might need a specific type of storage such as SSD.

A **node pool** represents a group of nodes in a GKE cluster that have the same configuration in terms of a specific CPU family, minimum CPU platform, preemptible VMs, or a specific storage requirement. A node pool is defined using a `nodeConfig` specification. All matching nodes that match the `nodeConfig` specification will be labeled using a node label where the key is `cloud.google.com/gke-nodepool` and the value is the name of the node pool.

The following is an example of a `nodeConfig` specification with a specific machine type, OAuth scopes, and a disk type:

```
nodeConfig: {
        machineType: "n2-highmem-32",
        oauthScopes: [
            "https://www.googleapis.com/auth/compute",
            "https://www.googleapis.com/auth/logging.write",
            "https://www.googleapis.com/auth/monitoring"
        ],
        diskType: "pd-ssd"
    }
```

A cluster is always created with a default node pool with a specific number of nodes and a specific machine type (along with other attributes). Additional custom node pools can be added based on their respective `nodeConfig` and workload requirements.

The following are some of the key characteristics of a node pool:

- A new node pool, by default, runs the latest stable Kubernetes version.
- The Kubernetes version on existing node pools can either be configured for auto-upgrade or can be manually upgraded.
- A node pool can be individually resized, upgraded, or deleted without impacting other node pools. Any change to the node pool impacts all nodes within the pool.

The following are a few CLI commands that can perform actions on a node pool. These commands can be executed on the cluster that was previously created in this chapter – `my-first-cluster`.

The following CLI command creates a node pool with a specific machine type:

```
gcloud container node-pools create my-high-mem-pool --machine-
type=n1-highmem-8 --cluster=my-first-cluster --num-nodes=2 -
zone=us-central1-c
```

The created node pool will be reflected on the GKE Console against the cluster (refer to *Figure 8.16*):

Figure 8.16 – New custom node pool – my-high-mem-pool created

The following are other CLI commands to resize a node pool, upgrade to a specific version, or delete a node pool:

```
# Resize node pool
gcloud container clusters resize my-first-cluster --node-
pool=my-high-mem-pool --num-nodes=1 -zone=us-central1-c
# Upgrading node pool to specific cluster version
gcloud container clusters upgrade my-first-cluster --cluster-
version="1.17.17-gke.3000" --node-pool=my-high-mem-cluster
--zone=us-central1-c
# Delete a node pool
gcloud container node-pools delete my-high-mem-pool
--cluster=my-first-cluster --zone=us-central1-c
```

Node pools in a regional or multi-zonal cluster are replicated to multiple zones. Additionally, the workload can be deployed to a specific node pool by explicitly specifying the node pool name using a nodeSelector or by finding a node pool that satisfies the resource requests as defined for the workload.

If the node pool name is explicitly specified using the nodeSelector attribute, then kube-scheduler will deploy workloads to the specified node. Otherwise, kube-scheduler will find the node pool that meets the intended resource request for the workload.

This completes the overview of GKE node pools. The next topic deep-dives into the various cluster configurations available in GKE.

GKE cluster configuration

GKE offers multiple cluster configuration choices based on cluster availability type, cluster version, network isolation, and Kubernetes features. Each of these configuration choices is discussed in the following sub-sections.

Cluster availability type

GKE allows you to create a cluster based on the availability requirements of the workloads. There are two types of cluster configuration based on availability types – zonal clusters (single-zone or multi-zonal) and regional clusters. These are discussed in the following sub-sections.

Zonal clusters

A **zonal cluster** will have a single control plane running in a single zone. The nodes (that is, worker nodes) can run either in a single zone or run across multiple zones. If the nodes run in the same zone as the control plane, then it represents a **single-zone cluster**. However, if nodes run across multiple zones, then it represents a **multi-zonal cluster**. Note that GKE allows up to 50 clusters per zone.

A multi-zonal cluster will only have a single replica of the control plane. The choice between a single zone or multi-zonal cluster is based on the level of availability required for an application. Specific to a multi-zonal cluster and in the event of a cluster upgrade or a zone outage, the workloads running on the nodes will continue to run, but a new node or workload cannot be configured till the cluster control plane is available.

The following are CLI commands to create a zonal cluster (single zone and multi zonal):

```
#Syntax
gcloud containers clusters create CLUSTER_NAME \
   --zone COMPUTE_ZONE \
   --node-locations COMPUTE_ZONE, COMPUTE_ZONE, [..]


#Single Zone Cluster
gcloud containers clusters create single-zone-cluster \
   --zone us-central1-a
#Multi Zonal Cluster
gcloud containers clusters create single-zone-cluster \
   --zone us-central1-a \
   --node-locations us-central1-a,us-central1-b, us-central1-c
```

The input parameter specific to the zone refers to the location of the control plane. The node locations refer to the locations of the worker node(s) and are not required for a single zone cluster as it will be the same as the master control plane.

This completes a brief overview of GKE zonal clusters. The next topic will provide an overview of GKE regional clusters.

Regional clusters

A regional cluster provides high availability both in terms of worker nodes as well as the control plane. A regional cluster has multiple replicas of the control plane running across multiple zones in a region. The worker nodes are also replicated across multiple zones and the worker nodes run in conjunction in the same zone as the control plane. A regional cluster cannot be converted into a zonal cluster.

The following is the CLI command to create a regional cluster:

```
#Syntax
gcloud containers clusters create CLUSTER_NAME \
   --region COMPUTE_REGION \
   --node-locations COMPUTE_ZONE, COMPUTE_ZONE, [..]
#Regional Cluster
gcloud containers clusters create single-zone-cluster \
   --region us-central1 \
   --node-locations us-central1-a,us-central1-b, us-central1-c
```

The input parameter specific to `region` refers to the location of the control plane. The node locations refer to the locations of the worker node. This is required for a multi-zone cluster as node locations could be in multiple zones.

This completes a brief overview of GKE cluster configuration based on cluster availability type. The next topic will provide an overview of GKE cluster configuration based on cluster version.

Cluster versions

GKE allows you to choose the cluster version. The cluster version can be a very specific version, the current default version, or can be based on a release channel, which is a combination of features based on early availability and stability. These cluster version configurations are discussed in the following sub-sections.

Specific versions

A GKE cluster can be created by specifying a specific version. This information can be provided as part of the *Static Version* selection while creating the cluster from the console. The user will be provided with a choice of cluster versions and can select an available version.

Release channels

Open source Kubernetes or K8s has a constant stream of releases. These could be required for the following purpose:

- To fix known issues

- To add new features

- To address any security risks/concerns

Kubernetes users who run applications on a Kubernetes cluster will prefer to exercise control in terms of how frequently the releases should be applied or the rate at which new features should be adopted. Google provides this choice to customers using the concept of a **release channel**.

Each of the release channels provides **generally available (GA)** features but the maturity of the features in terms of their original release date will vary from one channel to another. In addition, Google can also add the latest GKE-specific features depending on the type of release channel. This ensures that a specific feature or fix has potentially gone through the grind and is vetted in terms of its correctness and consistency over a period.

GKE provides three release channels:

- **Rapid**: This release channel includes the latest Kubernetes and GKE features when compared to other release channels, but the features are still several weeks old after their respective open source GA release.

- **Regular**: This is the default release channel, which includes Kubernetes and GKE-specific features that are reasonably new but are more stable in nature. The features are at least 2-3 months old after their release in the rapid channel and several months old from their open source GA release.

- **Stable**: This is the most stable of the release channels since the features added to this channel are added at least 2-3 months after being added to the regular channel. Essentially, the features are thoroughly validated and tested to provide the utmost stability.

The following is the CLI command to enroll a cluster in a release channel:

```
#Syntax
gcloud containers clusters create CLUSTER_NAME \
  --zone COMPUTE_ZONE \
  --release-channel CHANNEL \
  ADDITIONAL_FLAGS
# Release Channel Example
gcloud containers clusters create my-cluster \
  --zone us-central1-a \
  --release-channel rapid
```

To summarize, new Kubernetes versions and GKE features are promoted from the rapid to the regular to the stable channel, providing users with the choice to use newer features over stable features. GKE handles the availability of versions and the upgrade cadence once a cluster is added to the release channel. Each of the release channels continues to receive critical security updates.

The default version

If a specific version or a release channel is not specified, then GKE creates a cluster with the current default version. GKE selects a default version based on usage and real-world performance. GKE is responsible for changing the default version on a regular basis. Historically, new versions of Kubernetes are released every 3 months.

This completes a brief overview of GKE cluster configuration based on cluster version. The next topic will provide an overview of GKE cluster configuration based on network isolation choices.

Network isolation choices

There are two specific choices related to network isolation – a public cluster or a private cluster. A public cluster is the default configuration. However, this does not enforce network isolation and the cluster is accessible from any public endpoint. This makes the cluster vulnerable from a security standpoint. The drawbacks of configuring a public cluster can be handled through a private cluster, which is introduced in the following sub-sections.

Private clusters

GKE provides an option to create a private cluster where the nodes only have internal IP addresses. This means that the nodes and the pods running on the nodes are isolated from the internet and inherently will not have inbound or outbound connectivity to the public internet.

A private cluster will have a control plane that includes a private endpoint, in addition to a public endpoint. Access to the public endpoint can be controlled through multiple options. In addition, the control plane will run on a VM that is in a VPC network in a Google-owned project. The details surrounding private clusters will be discussed in depth as part of *Chapter 9, Securing the Cluster Using GKE Security Constructs*.

Kubernetes features – alpha clusters

New features in Kubernetes are rolled out to GKE as part of the release channel in most cases. The release channel includes choices of rapid, regular, and stable. However, alpha features are only available in special GKE alpha clusters. This is discussed in the following sub-sections.

Alpha clusters

Alpha clusters are a specific feature of GKE that is designed for adopting new features that are not production-ready or generally available as open source. GKE creates alpha clusters as short-lived clusters and they are automatically deleted after 30 days.

The following is the CLI command to create an alpha cluster:

```
#Syntax
gcloud container clusters create cluster-name \
    --enable-kubernetes-alpha \
    [--zone compute-zone] \
    [--cluster-version version]
#Alpha Cluster Example
gcloud container clusters create my-cluster \
    --enable-kubernetes-alpha \
    --region us-central1
```

These clusters do not receive security updates, cannot be auto-upgraded or auto-repaired, and are not covered by any GKE-specific SLAs. Hence, alpha clusters are never recommended for production workloads.

This completes a brief overview of GKE cluster configuration based on network isolation choices. This also concludes the sub-section on GKE cluster configuration in general. The next topic details possible autoscaling options in GKE.

AutoScaling in GKE

There are three potential options to perform autoscaling in GKE. Each of these options is suitable for specific needs and situations:

- **Cluster autoscaler**: A scaling option to resize a node pool in a GKE cluster

- **Horizontal Pod Autoscaler** (**HPA**): An option that indicates when application instances should be autoscaled based on their current utilization

- **Vertical Pod Autoscaler** (**VPA**): An option that suggests recommended resources for a Pod based on the current utilization

The upcoming topics detail the preceding autoscaling mechanisms, starting with the cluster autoscaler.

The cluster autoscaler

The **cluster autoscaler** is a scaling mechanism to automatically resize a node pool in a GKE cluster. The scaling is based on the demands of workloads deployed within the node pool. This allows you to implement the core concept of cloud computing, called elasticity, and removes the need to over-provision or under-provision nodes.

The cluster autoscaler works on a per-node pool basis and is based on resource requests (defined as part of the Pod specification) rather than the actual resource utilization. When a new Pod needs to be deployed, the Kubernetes scheduler works out of the Pod resource requests and attempts to find a node to deploy the Pod. If there is no node that matches the Pod resource requirement in terms of available capacity, then the Pod goes into a pending state until any of the existing pods are terminated or a new node is added.

The cluster autoscaler keeps track of the pods that are in the pending state and subsequently tries to scale up the number of nodes. Similarly, the cluster autoscaler also scales down the number of nodes if the nodes are under-utilized. A minimum or maximum number of nodes can be defined for the cluster autoscaler, which allows it to operate within the specified limits.

When a cluster is scaled down, there is a possibility that new workloads might have to wait till new nodes are added. This could cause a potential disruption. GKE profile types provide a choice of options to choose between balanced and aggressive scale-down:

- **Balanced**: The default profile option, which is not aggressive in nature.

- **Optimize-utilization**: Scaling down is more aggressive and removes underutilized nodes faster.

The following are some CLI commands related to the cluster autoscaler:

```
# Create cluster with autoscaler limits
gcloud container clusters create my-autoscaler-cluster \
    --zone us-central1-b \
    --num-nodes 3 --enable-autoscaling --min-nodes 1 --max-nodes
5
# Update autoscaling profile to optimize-utilization
gcloud beta container clusters update my-autoscaler-cluster \
    -autoscaling-profile optimize-utilization
```

The following are some limitations that need to be considered when using the cluster autoscaler:

- There is a graceful termination of 10 minutes for rescheduling pods on to a different node before forcibly terminating the original node.

- The node pool scaling limits are determined by zone availability. If a cluster has 3 nodes (with min_nodes = 1 and max_nodes = 5) across 4 zones, then if 1 of the zones fails, the size of the cluster can vary from 4-20 nodes per cluster to 3-15 nodes per cluster.

This concludes the overview of the cluster autoscaler. The next topic focuses on the **Horizontal Pod Autoscaler (HPA)**.

The Horizontal Pod Autoscaler

The HPA is a Kubernetes controller object that automatically scales the number of pods in a replication controller, Deployment, ReplicaSet, or StatefulSet based on the observed CPU or memory utilization. The HPA indicates the Deployment or StatefulSet against which scaling needs to happen. The HPA doesn't apply to DaemonSets.

To implement the HPA, the following factors need to be considered:

- One HPA object needs to be defined per Deployment or StatefulSet.

- The attribute `--horizontal-pod-autoscaler-sync-period` allows you to implement the HPA as a control loop. The default value is 15 seconds per period.

- `kube-controller-manager` (on a per-period basis) obtains metrics from the resource manager API or the custom metrics API and compares them against the metrics specified in each HPA definition.

The following are few key parameters that can define the HPA configuration:

- `--horizontal-pod-autoscaler-initial-readiness-delay`: A configurable window to ensure that a Pod is transitioned to the ready state.

- `--horizontal-pod-autoscaler-cpu-initialization-period`: A configurable window to set the CPU initialization period, once the Pod is transitioned to the ready state. The default is 5 minutes.

- `--horizontal-pod-autoscaler-downscale-stabilization`: A configurable window that autoscaler needs to wait before initiating a downscale operation after the current one is completed. The default is 5 minutes. This prevents thrashing.

The following is the sample definition of an HPA object based on CPU utilization:

```
apiVersion: autoscaling/v1
kind: HorizontalPodAutoscaler
metadata:
  name: nginx
spec:
  scaleTargetRef:
    apiVersion: apps/v1
    kind: Deployment
    name: my-nginx
  minReplicas: 1
  maxReplicas: 5
  targetCPUUtilizationPercentage: 75
```

In the preceding example, `kube-controller-manager` will scale up the Deployment based on the HPA object specification, to a maximum of 5 instances if the target CPU utilization exceeds 75%. This concludes the overview of the HPA. The next topic focuses on the **Vertical Pod Autoscaler (VPA)**.

The Vertical Pod Autoscaler (VPA)

The cluster autoscaler functions based on the workload's CPU and memory request limits. If these limits are not defined appropriately, then there is always a chance of over-provisioning or under-provisioning as the reference values will not be accurate.

The VPA is a Kubernetes resource that recommends values for CPU and memory requests/limits. Additionally, the VPA can automatically update workloads if the `updateMode` attribute is set to *On* on the VPA. This will potentially evict the existing Pod as a change is required to the pod's resource requests and will result in a new Pod with the updated recommendations.

This ensures that the cluster nodes are optimally utilized and potentially removes the need to run benchmark tests to determine the correct values for CPU and memory requests. VPA communicates with the cluster autoscaler to perform the appropriate operations on the nodes tied to the node pools.

The following is a sample definition of a VPA object:

```
apiVersion: autoscaling.k8s.io/v1
kind: VerticalPodAutoscaler
metadata:
  name: my-vpa
spec:
  targetRef:
    apiVersion: "apps/v1"
    kind:       Deployment
    name:       my-nginx
  updatePolicy:
    updateMode: "On"
```

The `kind` attribute in the preceding snippet indicates that the Kubernetes resource is a VPA object. The `updateMode` attribute indicates that the recommendations suggested by the VPA are automatically applied against the running workloads.

The following are some CLI commands specific to the VPA:

```
# To view recommendations of VPA is updateMode was set to Off
kubectl get vpa my-vpa --output yaml
# To disable VPA
gcloud container clusters update my-cluster --no-enable-
vertical-pod-autoscaling
```

If an HPA object is configured to evaluate metrics for CPU or memory, it's recommended that HPA should not be used with VPA.

> **Multi-dimensional Pod autoscaling (MPA)**
>
> This is a new autoscaling option that is currently in pre-GA. As per this option, it is possible to configure autoscaling to horizontally scale based on CPU and vertically scale based on memory at the same time. MPA is supported for clusters that are 1.19.4-gke.1700 or later.

This concludes the section on autoscaling in GKE where multiple mechanisms were detailed out. The next section focuses on networking constructs with respect to GKE. This will cover details about Pod networking, Service networking, and will deep dive into the usage of GKE load balancers to expose services for external consumption.

Networking in GKE

Applications are deployed in Kubernetes as containers. Pods run containers. The desired state of the pods is controlled by Deployments and the applications are exposed for both internal and external networking through Services. The deployed pods run in GKE on nodes. Nodes in GKE are represented by virtual machines or VMs. These nodes are deployed in a **Virtual Private Cloud** (**VPC**).

A VPC defines a virtual network topology that closely resembles a traditional network. It is a logically isolated network and provides connectivity between deployed resources. A VPC also provides complete control in terms of launching resources, selecting a range of RFC 1918 addressing, the creation of subnets, and so on.

A VPC on GCP has a pre-allocated IP subnet, for every GCP region. When a GKE cluster is deployed within the VPC, a specific region or zone can be selected. Since GKE nodes are made up of Compute Engine VMs and these VMs need an IP address, the range of IP addresses is allocated from the IP subnet pre-allocated to the region. A VPC on GCP is considered a global resource since a single Google Cloud VPC can span multiple regions without communicating across the public internet. It is not required to have a connection in every region.

GCP provides the option of configuring alias IP ranges. This allows VMs to have an additional secondary IP address. As a result, a VM can have multiple services running with a separate IP address. These secondary IP addresses are routable within the VPC without the need to configure additional routes.

A GKE cluster might need to run cluster-wide services. GCP recommends deploying a GKE cluster as a **VPC-native cluster**. A VPC-native cluster uses three unique subnet IP address ranges:

- A primary IP address range of subnet for node IP addresses

- A secondary IP address range for all Pod IP addresses

- An additional secondary IP address range for all Service IP addresses

GKE provides flexibility where the number of nodes in a cluster and the maximum number of pods per node are configurable. The next topic details how pods are assigned IP addresses when pods are deployed in a GKE cluster.

Pod networking

When a Pod is scheduled on a node, Kubernetes creates a network namespace for the Pod on the node's Linux kernel and connects the node's physical network interface to the Pod with a virtual network interface, thus allowing communication among pods within the same node.

Kubernetes assigns an IP address (the Pod's IP) to the virtual network interface in the Pod's network namespace from a range of addresses reserved for Pods on the node. This address range is a subset of the IP address range assigned to the cluster for Pods, which can be configured when creating a cluster.

GKE automatically configures VPC to recognize this range of IP addresses as an authorized secondary subnet of IP addresses. As a result, the pod's traffic is permitted to pass the anti-spoofing filters on the network. Also, because each node maintains a separate IP address base for its pods, the nodes don't need to perform network address translation on the pod's IP address. The next topic details Service networking, specifically, how services can effectively receive traffic from external sources via the use of GKE load balancers.

Service networking

A Service is a Kubernetes resource that creates a dynamic collection of IP addresses called endpoints. These IP addresses belong to the Pod that matches the Service label selector. Kubernetes creates a Service by assigning a static virtual IP address and this IP address is assigned from the pool of IP addresses reserved for services by the cluster.

Out of the available Service types, the `LoadBalancer` Service type is implemented in GKE using GCP's **Network Load Balancer** (**NLB**). The NLB supports TCP and UDP traffic. GCP creates a network load balancer when a Service of type `LoadBalancer` is created within the GKE cluster. GCP will subsequently assign a static `LoadBalancer` IP address that is accessible from outside the cluster and the project.

For traffic sent to the GCP NLB, *Figure 8.17* depicts the interactions between the NLB and the nodes within the GKE cluster. These interactions are listed as follows in a step-by-step manner:

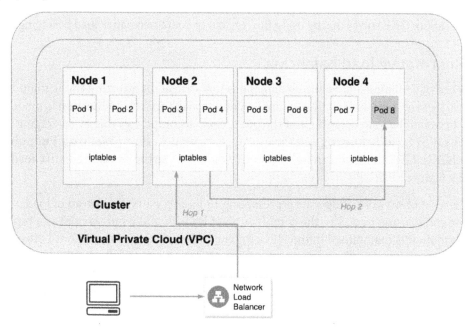

Figure 8.17 – Interactions between the NLB and a GKE cluster within a VPC

Step-by-step interactions:

1. NLB will pick a random node in the cluster and forwards the traffic (say **Node 2** as per *Figure 8.17*)

2. The Service might be tied to multiple pods spread across the cluster nodes. The kube-proxy Service on the node receives the client request and will select a Pod matching the Service at random. The selected Pod can be on the same node or a different node.

3. If the selected Pod is on a different node (say **Pod 8**), then the client request will be sent to the other node (**Node 4**) from the original node (**Node 2**). The response goes back to the original node (**Node 2**) that received the request and subsequently goes back to the client.

The preceding process provides a way to access services from an external client and maintains an even balance with respect to Pod usage. However, there is a possibility that within the Kubernetes cluster, a response might have to go through multiple nodes as the request was directed from one node to the other, resulting in a **double hop**.

To avoid **double hop**, Kubernetes natively provides an option called
`externalTrafficPolicy`. If set to local, `kube-proxy` will pick a Pod on the local node
(either **Pod 3** or **Pod 4**) and will not forward the client request to another node. However,
this creates an imbalance and users must choose between better balance versus low-latency
communication. GKE solves this by using the concept of container-native load balancing.

Container-native load balancing

The essence of container-native load balancing is that instead of directing traffic to nodes,
traffic will be sent to pods directly, avoiding an additional hop. The connection is made
directly between the load balancer and the pods. GKE accomplishes this by leveraging
GCP HTTP(S) Load Balancing and the use of a data model called a **Network Endpoint
Group** (**NEG**). GKE needs to run in VPC-native mode to use the container-native load
balancing feature.

A NEG is a set of network endpoints representing IP to port pairs. So instead of load
balancing traffic using node IPs, the combination of Pod IPs and a port is used as a tuple.
This information is maintained in the NEG. *Figure 8.18* depicts the interactions between
GKE container-native load balancing and pods in GKE nodes through an NEG. As per
Figure 8.18, a request to the container-native load balancer is forwarded to the NEG. The
NEG then chooses the specific Pod based on the request, and directly forwards the traffic
to the node associated with the Pod in a single hop, thus avoiding the *double hop*:

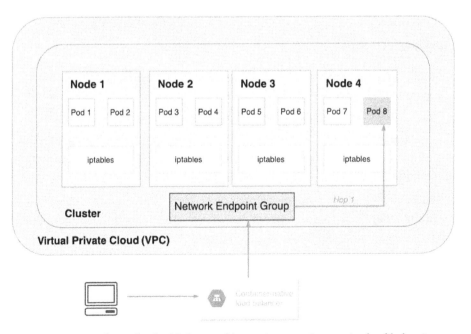

Figure 8.18 – Solving the double hop problem using container-native load balancing

Apart from establishing a direct connection to the Pod, container-native load balancing allows direct visibility of Pods, leading to the possibility of accurate health checks. The source IP address is preserved thus giving insights into the roundtrip time between the client and the load balancer.

This concludes a high-level overview of networking constructs specific to GKE. The next section summarizes the storage options available for containerized applications deployed in GKE.

Storage options for GKE

Kubernetes offers storage abstractions in the form of Volumes and Persistent Volumes. These are used as storage options providing file system capacity that is directly accessible by applications running in a Kubernetes cluster. Persistent Volumes exist beyond the life of a container and can further be used as durable file storage or as a database backing store.

In GKE, Compute Engine persistent disks are used as persistent volumes. GKE also provides various managed backing stores such as Cloud SQL, Cloud Datastore, and so on, which removes the need to run a database as an application inside the GKE cluster, connecting applications in a GKE cluster to a managed datastore instead. For example, a frontend application in a GKE Cluster can be connected to Cloud SQL rather than the frontend application connecting to another application running a MySQL server. To be more specific, the frontend application can connect to Cloud SQL for database needs through a Cloud SQL proxy. This can be run inside the frontend application's Pod as a side-car container.

This abstracts away infrastructure requirements and reduces maintenance, allowing you to focus on the application. GCP offers managed services across relational, non-relational, and caching services that applications running in a GKE cluster can connect to.

In addition to applications that might require a backend data store, there could be applications running in a GKE cluster that might need object storage. **Google Cloud Storage** (**GCS**) is an object storage Service. Object-based storage refers to the storage of an ordered group of bytes where the structure and semantics of those bytes are not important. It can be used for a variety of applications, such as the following:

- Serving images for a website
- Streaming music, videos, and media hosting
- Constructing data lakes for analytics and machine learning workloads

Applications within the GKE cluster can access Cloud Storage using Cloud Storage APIs. This concludes the summary of the storage options available in GCP for applications deployed in GKE. The next section summarizes details on cloud operations from a GKE perspective.

Cloud Operations for GKE

Google Kubernetes Engine (**GKE**) provides native integration with **Google's Cloud operations** – a suite of tools that allows you to monitor workloads, collect application logs, capture metrics and provide alerting or notification options on key metrics. Cloud operations and the respective suite of services are elaborated on in detail as part of *Chapter 10*, *Exploring GCP Cloud Operations*.

Cloud Operations for GKE is enabled by default at the time of cluster creation. However, it is possible to configure if the user chooses to disable Cloud Monitoring or Cloud Logging as part of the GKE cluster configuration. Cloud Operations for GKE monitors GKE clusters and provides a tailored, out-of-the-box dashboard that includes the following capabilities:

- Viewing cluster resources categorized by infrastructure, workloads, or services

- Inspecting namespaces, nodes, workloads, services, pods, and containers

- Viewing application logs for pods and containers

- Viewing key metrics related to clusters, such as CPU utilization, memory utilization, and so on

Logging and monitoring are two critical aspects of reliably running a Service or application in a GKE cluster. These will be covered as part of upcoming topics from the aspect of Cloud Operations for GKE.

Logging for GKE

GKE deploys applications and orchestrates multiple actions or events within a cluster. This results in a variety of logs such as application logs, system logs, event logs, and so on. Logging provides visibility of various actions that happen and is also considered a passive form of monitoring.

There are two options to view logs for a GKE cluster:

- Kubernetes Native Logging
- GKE Cloud Logging

Kubernetes Native Logging

Kubernetes supports native logging to standard output and standard error. In Kubernetes, the *container engine* can be used to redirect stdin/out and standard error streams from the containers to a logging driver. This driver is configured to write these container logs in JSON format and store them in the `/var/log` directory at the node level. This includes logs from containers and logs from node control plane components such as `kubelet` and `kube-proxy`. These logs can be retrieved using the `kubectl logs` command.

The `kubectl logs` command can be used to retrieve logs for a Pod or a specific container within a Pod. The command also provides options to retrieve logs for a specific period or you can retrieve a portion of logs using the `tail` option. A few of such examples are provided as follows:

```
# Stdout container logs; pod has a single container
kubectl logs <pod-name>


# Stdout container logs; pod has multiple containers
kubectl logs <pod-name> -c <container-name>


# Stdout container logs - most recent 50 lines
kubectl logs --tail=50 <pod-name>


# Stdout most recent container logs in the last 1 hour
kubectl logs --since=1h <pod-name>


# Stream pod logs
kubectl logs -f <pod-name>
```

Kubernetes native logging can lead to node saturation as the log files continue to grow in the node's storage directory. GKE solves this to an extent by running the Linux log rotate utility to clean up the log files. Any log files older than a day or more than 100 MB will be automatically compressed and copied into an archive file.

GKE only stores the five most recently archived log files on the nodes and will delete the previous archived log files. Though this ensures that the node doesn't saturate in terms of disk space, it still poses a problem if older application logs need to be analyzed or researched.

By default, open source Kubernetes or K8s will delete logs related to a container either when a container is deleted or when a Pod tied to the container is deleted. GKE resolves problems related to node saturation and provides the ability to analyze logs related to deleted pods/containers by streaming the logs to Cloud Logging, as part of Cloud Operations. Application logs, system logs, and log events can be streamed to Cloud Logging, which will be discussed as part of upcoming topics.

GKE Cloud Logging

Open source Kubernetes or K8s doesn't provide a log storage solution for cluster-level logging. GKE handles this by streaming log events to Cloud Logging. **Cloud Logging** is a centralized log management utility and a fully managed Service. Cloud Logging can automatically scale and can ingest terabytes of log data per second.

GKE streams to Cloud Logging by using `FluentD` logging agents. A `FluentD` agent is implemented as a DaemonSet because it needs to run on every node in the cluster.

Logging agents are pre-installed on each node as a DaemonSet and are pre-configured to push log data to Cloud Logging. `FluentD` collects container logs and system logs from the node. FluentD aggregates the logs, appends additional metadata, and pushes them to Cloud Logging.

Figure 8.19 illustrates the interactions of logs being sent from GKE to Cloud Logging using the `FluentD` DaemonSet Pod on each node in the cluster:

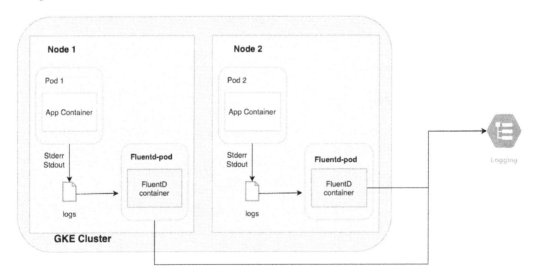

Figure 8.19 – FluentD agent capturing logs and sending to Cloud Logging

Event logs are also streamed to Cloud Logging. Event logs refers to logs from operations that take place on the cluster such as the creation/deletion of a Pod, scaling of deployments, and so on. Events are stored as API objects on the Kubernetes master or control plane. GKE uses an event exporter in the cluster master to capture the events and automatically pushes them to Cloud Logging.

Cloud Logging provides the ability to capture metrics from streaming logs and create alerting policies as needed. Cluster actions such as autoscaling can be configured based on custom metrics. By default, GKE-specific logs related to a cluster are available in Cloud Logging for 30 days. For longer retention, Cloud Logging offers options to export logs to Cloud Storage or Big Query using the concept of log sinks. *Chapter 10, Exploring GCP Cloud Operations*, elaborates on topics related to Cloud Logging in depth.

Monitoring for GKE

Monitoring provides insights into how an application or Service functions based on key internal metrics related to a GKE cluster. In addition, monitoring also provides insights from a user's perspective based on the user's interaction with the Service. The previous chapters on site reliability engineering (*Chapter 1, DevOps, SRE, and Google Cloud Services for CI/CD*, to *Chapter 4, Building SRE Teams and Applying Cultural Practices*), clearly call out Service reliability as one of the key aspects. Monitoring is the fundamental input to ensure that a Service runs reliably.

Monitoring provides data that is critical to make decisions about applications. This data can be used further to resolve an ongoing incident and perform a blameless postmortem, and you can use it further to improve an existing test suite and provide inputs to the product and development team for any further improvements or fine-tuning.

Cloud Monitoring is Google's managed solution that provides a solution to monitor the state of services using key parameters such as latency, throughput, and so on, and identify performance bottlenecks. From a GKE perspective, monitoring can be divided into two domains:

- **Cluster-Level Monitoring**: This includes monitoring cluster-level components such as nodes and components from the master control plane such as `kube-apiserver`, `etcd`, and other infrastructure elements.

- **Pod-Level Monitoring**: This includes monitoring resources using container-specific metrics, tracking deployment-specific system metrics, tracking instances, monitoring uptime checks, and monitoring application-specific metrics designed by the application's developer(s).

Kubernetes uses the concept of labels to group or track resources. The same concept can be extended, and resources can be filtered in Cloud Monitoring using labels. Cloud Monitoring provides ways to track all relevant metrics and put them on a customized dashboard, thus giving visibility of a GKE cluster. *Figure 8.20* shows the built-in **GKE Dashboard** from Cloud Monitoring (with options displayed in collapsed mode). The GKE dashboard summarizes information about clusters, namespaces, nodes, workloads, Kubernetes services, Pods, and Containers:

Figure 8.20 – Built-in GKE Dashboard from Cloud Monitoring

This completes the topic on Cloud Operations for GKE and concludes the section on GKE where many key concepts and core features were discussed in detail. The next section elaborates on the latest operation mode in GKE, called **Autopilot**.

GKE Autopilot – hands-on lab

GKE Autopilot or Autopilot is one of the two modes of operation supported by GKE. The other mode being the standard mode (which was elaborated on at the start of this chapter). Autopilot removes the need to perform **do-it-yourself** (**DIY**) actions during cluster creation and instead creates a cluster with the industry-standard recommendations regarding networking and security. In addition, Autopilot removes the need to configure node pools or estimate the size of the cluster upfront. Nodes are automatically provisioned based on the types of deployed workloads and the user is essentially charged for the running workloads.

Autopilot is not only managed but is also a serverless K8s offering from GKE. Autopilot, however, does not offer all cluster configuration choices offered by the standard mode. The following table represents the configuration choices offered by Autopilot in comparison to the standard mode:

Cluster Choice	Autopilot Mode	Standard Mode
Availability Type	Regional	Regional, Zonal
Version	Release Channel	Release Channel, Default, Specific
Network Isolation	Private or Public	Private or Public
Kubernetes Features	Production	Production or Alpha

The following is a step-by-step guide to creating a GKE cluster in Autopilot mode:

1. Navigate to the GCP Console and select the compute Service – **Kubernetes Engine**.

2. Select the option to create a cluster and choose **Autopilot** mode. Refer to *Figure 8.21*:

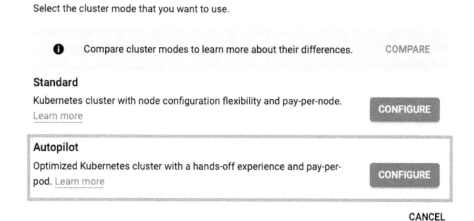

Figure 8.21 – Select Autopilot mode during cluster creation

3. Enter the name for the cluster as `my-autopilot-cluster`. Leave the default selections for the rest of the options and select the **CREATE** action. Refer to *Figure 8.22*:

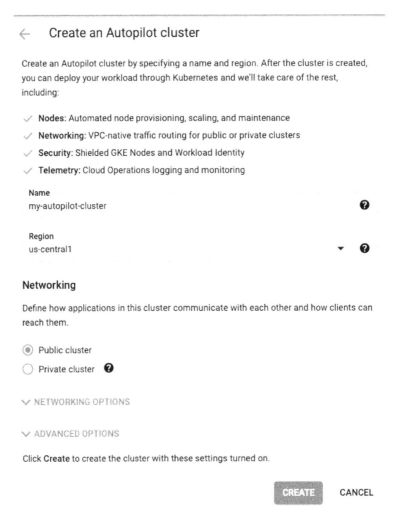

Figure 8.22 – Creating a cluster in Autopilot mode

4. This will initiate the cluster creation process but in Autopilot mode. Once the cluster is created, the cluster will be listed on the cluster list page as shown in *Figure 8.23*:

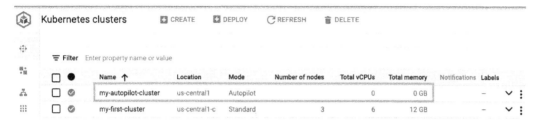

Figure 8.23 – New cluster created in Autopilot mode

Here are some observations from the newly created Autopilot cluster. These observations differentiate the Autopilot cluster from a Standard mode cluster:

- An autopilot cluster is created without pre-assigning any nodes upfront.

- An autopilot cluster is always created as a regional cluster.

- The release channel for an autopilot cluster is the *Regular channel*.

- Node auto-provisioning and vertical Pod autoscaling are enabled by default.

- Advanced networking options such as intranode visibility, NodeLocal DNSCache, and HTTP load balancing are enabled by default.

- Security options such as Workload Identity and shielded GKE nodes are enabled by default. These security options are discussed in *Chapter 9, Securing the Cluster Using GKE Security Constructs*.

Once a cluster is created in Autopilot mode, workloads can be deployed to the Autopilot cluster in the exact same way that workloads were previously deployed to a cluster in Standard mode. *Figure 8.24* refers to a Deployment created on the Autopilot cluster:

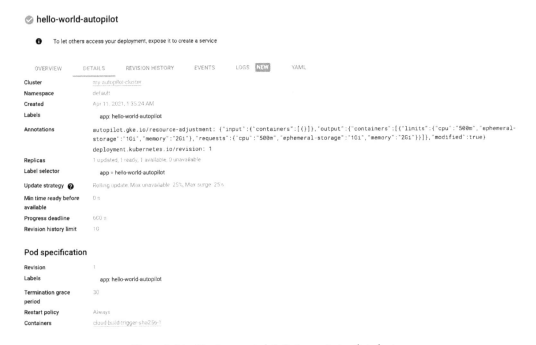

Figure 8.24 – Deployment details in an Autopilot cluster

The resources required to run the workloads are allocated to the Autopilot cluster. *Figure 8.25* displays the cluster list page with resources allocated to `my-autopilot-cluster`. In this specific case, 0.5 vCPUs and 2 GB memory are allocated to run a single Pod. So, the user is only charged for this workload:

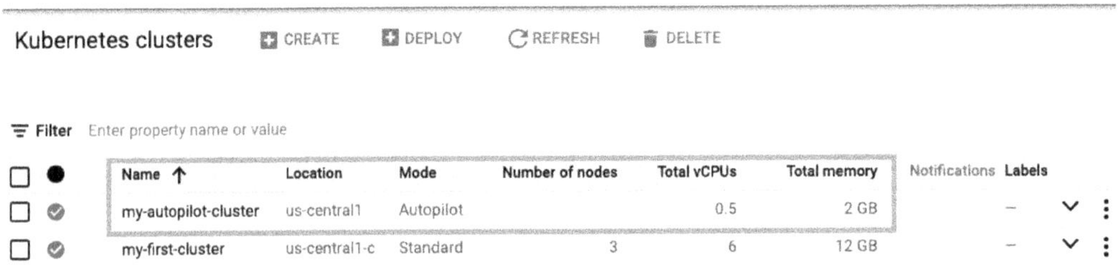

Figure 8.25 – Resource allocation for the Autopilot cluster after deploying a workload

This completes the hands-on lab related to GKE Autopilot. This lab provides insights into the Autopilot configuration and how resources are allocated to the cluster after the deployment of workloads. This also brings us to the end of the chapter.

Summary

Given that open source Kubernetes or K8s involves a lot of setup and upkeep, we deep-dived into Google Kubernetes Engine or GKE, a GCP compute Service that runs containerized applications. The Kubernetes concepts learned in *Chapter 7, Understanding Kubernetes Essentials to Deploy Containerized Applications*, apply to GKE. We additionally explored GKE core features such as GKE node pools, GKE cluster configurations, autoscaling, and GKE's ability to integrate with other GCP services across networking and operations. The next chapter focuses on security-specific features related to the Google Kubernetes Engine, with the goal of hardening a cluster's security.

Points to remember

The following are some important points to remember:

- GKE is fully managed, uses a container-optimized OS, and supports autoscaling, the auto-repair of nodes, and auto-upgrades.

- GKE supports two modes of operations – Standard and Autopilot.

- GKE Standard mode supports VPC-native traffic routing and HTTP load balancing as default options.

- Cloud operations for GKE are enabled as a default setting.

- A private Kubernetes engine cluster cannot be accessed publicly.

- A node pool represents a group of nodes with the same configuration.

- By default, a new node pool runs the latest Kubernetes version and can be configured for auto-upgrade or can be manually upgraded.

- Node pools in a regional or multi-zonal cluster are replicated to multiple zones.

- A multi-zonal cluster will only have a single replica of the control plane.

- A regional cluster has multiple replicas of the control plane running across multiple zones in a region.

- A release channel is used to fix known issues or add new features or address any security risks or concerns.

- GKE creates a cluster with the default version if a specific version or release channel is not specified.

- Alpha features are only available in special GKE alpha clusters and are not available as part of release channels.

- Options to autoscale in GKE include the cluster autoscaler, HPA, VPA, and MPA (pre-GA).

- The cluster autoscaler automatically resizes a node pool in a GKE cluster.

- The HPA indicates when application instances should be scaled based on the current utilization.

- The HPA is not supported for DaemonSets.

- The VPA suggests recommended resources for a Pod based on the current utilization.

- The VPA can automatically update workloads if the `updateMode` attribute is set to *On*.

- MPA allows you to horizontally scale based on CPU and vertically scale based on memory at the same time. This is a pre-GA feature.

- Autoscaler provides two profile options to scale down: balanced and optimize-utilization.

- Kubernetes' native option to avoid double hop is to set `externalTrafficPolicy` to `local`.

- GKE avoids double hop using GCP HTTP(S) Load Balancer and an NEG.

- An NEG is a set of network endpoints representing IP to port pairs.

- GKE runs Linux's log rotate utility to clean up log files. Any log files older than a day or more than 100 MB will be automatically compressed and copied into an archive file.

- GKE only stores the five most recently archived log files on the nodes and will delete the previously archived log files.

- GKE streams to Cloud Logging by using FluentD logging agents.

- Event logs refers to logs from operations that take place on a cluster.

- Events are stored as API objects on the cluster master. GKE uses an event exporter to push events to Cloud Logging.

- GKE cluster-specific logs are available in Cloud Logging for 30 days.

- For longer retention, Cloud Logging can export logs using log sinks.

- GKE Autopilot mode supports cluster configurations where the availability type is *Regional*, the version is *Release Channel*, network isolation is *Private* or *Public*, and Kubernetes features are *Production*.

Further reading

For more information on GCP's approach to DevOps, read the following articles:

- **Kubernetes**: `https://kubernetes.io/docs/home/`
- **Google Kubernetes Engine**: `https://cloud.google.com/kubernetes-engine`

Practice test

Answer the following questions:

1. How do you create control plane components in GKE?

   a) Create worker nodes and then create control plane components on the worker nodes.

   b) A GKE cluster does not mandate the creation of control plane components.

   c) Create control plane components on a node group called `master` and the worker nodes are placed in a node group called `worker`.

   d) The control plane components are automatically created and managed by GKE on behalf of the user.

2. Pod `p1` has three containers – `c1`, `c2`, and `c3`. The user wants to view the logs of container `c2`. Select the option that represents the appropriate CLI command to view the logs:

   a) `kubectl logs -p p1 -c c2`

   b) `kubectl logs p1 -c c2`

   c) `kubectl logs pod=p1 container=c2`

   d) `kubectl logs p1 container=c2`

3. The company *Alpha* is about to launch a stateless web application to offer a new e-commerce Service. The web application will have steady traffic with occasional peaks, especially when special offers are announced for customers. Select the option that depicts an appropriate cluster design in this case:

   a) Deploy a standard cluster and use a Deployment with the HPA.

   b) Deploy a cluster with autoscaling and use a Deployment with the HPA.

   c) Deploy a standard cluster and use a Deployment with the VPA.

   d) Deploy a cluster with autoscaling and use a Deployment with the VPA.

4. Choose the cluster configuration that could withstand it if there was a loss of a GCP zone:

   a) Create a regional cluster.

   b) Create a Redis cluster that can cache the resource information of the zone where cluster resources are hosted.

   c) Create two clusters in separate zones and create a load balancer between them.

   d) None of the above.

5. Select the Google Cloud Service where private GKE clusters can use Docker images from?

   a) Cloud Source Repositories

   b) Container Registry

   c) Cloud Build

   d) All of the above

6. Select the allowed maximum clusters per zone:

   a) 25

   b) 50

   c) 100

   d) Unlimited

7. Select the command to get authentication credentials to interact with a cluster named my-cluster:

   a) `gcloud containers clusters get-credentials my-cluster`

   b) `gcloud container clusters get-credentials my-cluster`

   c) `gcloud container cluster get-credentials my-cluster`

   d) `gcloud containers cluster get-credentials my-cluster`

8. Select the command that can retrieve pods in a cluster:

   a) `gcloud get pods`

   b) `kubectl list pods`

   c) `gcloud list pods`

   d) `kubectl get pods`

9. The company *Real World* decides to use a third-party monitoring solution to monitor an application deployed in a GKE cluster. Select the best approach to deploy the third-party monitoring solution:

    a) It is not possible to use a third-party monitoring solution in GKE.

    b) Download the monitoring solution for Cloud Marketplace.

    c) Deploy the monitoring solution in a Pod as a DaemonSet.

    d) Deploy the monitoring solution in a Pod as a ReplicaSet.

10. A VPC on Google Cloud is a:

    a) Zonal resource

    b) Global resource

    c) Regional resource

    d) Multi-Regional resource

11. An application called *my-app* in GKE needs access to a managed MySQL database. Select the most appropriate option:

    a) Run MySQL as an application in the cluster. The *my-app* application will connect with the MySQL application through the ClusterIP Service.

    b) Use Cloud SQL to run MySQL database. Run the Cloud SQL proxy as a side-car container insider the application's Pod.

    c) Run MySQL as an application in the cluster. The *my-app* application will connect with the MySQL application through the LoadBalancer Service.

    d) Use Cloud SQL for running MySQL Database. Run the Cloud SQL proxy as a ClusterIP Service.

12. Google Network Load Balancing distributes the following traffic:

    a) TCP

    b) UDP

    c) TCP or UDP

    d) None of the above

13. From an availability-type point of view, a cluster created in *Autopilot* mode is:

    a) Zonal

    b) Multi-zonal

    c) Regional

    d) Zonal and regional

14. Select the option that is not a supported release channel in GKE:

    a) Regular

    b) Alpha

    c) Rapid

    d) Stable

15. Select the possible cluster configurations based on network isolation:

    a) Standard and Private

    b) Standard and Public

    c) Standard and Default

    d) Private and Public

Answers

1. (d) – The control plane components such as the `kube-api` server, scheduler, and so on form the cluster master and are set up and managed by GKE.

2. (b) – `kubectl logs p1 -c c2`

3. (b) – Deploy a cluster with autoscaling and use Deployment with HPA.

4. (a) – Create a regional cluster as the workload is spread across multiple zones in one region.

5. (b) – Container Registry

6. (b) – 50

7. (b) – `gcloud container clusters get-credentials my-cluster`

8. (d) – `kubectl get pods`

9. (c) - Deploy the monitoring solution in a Pod as a DaemonSet.

10. (b) – Global resource

11. (c)

12. (c) – TCP or UDP

13. (c) – Regional

14. (b) – Alpha

15. (d) – Private and Public

9
Securing the Cluster Using GKE Security Constructs

Kubernetes, or K8s, is an open source container orchestration system that runs containerized applications but requires significant effort to set up and maintain. **Google Kubernetes Engine (GKE)** is an enhanced version of K8s that is managed in nature, abstracts the master plane components from the user, provides the ability to auto-upgrade, and supports features such as DNS, logging, and monitoring dashboards as built-ins rather than maintaining them as external plugins. Kubernetes has a lot of critical concepts, jargon, and objects. The last two chapters (*Chapter 7, Understanding Kubernetes Essentials to Deploy Containerized Applications*, and *Chapter 8, Understanding GKE Essentials to Deploy Containerized Applications*) focused on native Kubernetes features such as cluster anatomy, elaborated on key Kubernetes objects, and discussed how applications are scheduled on a cluster. In addition, the focus was extended to learning about specific GKE features such as node pools, cluster configurations, options to auto scale workloads, and understand how GKE interacts with other GCP services.

This chapter specifically focuses on understanding the basic security constructs in Kubernetes, their application, and then specific GKE security features that are fundamental to hardening a cluster's security. The key here is to secure the applications running inside the cluster using GKE-specific features.

This chapter will cover the following topics:

- **Essential security patterns in Kubernetes**: This section deep dives into fundamental security constructs in native Kubernetes, such as authentication, authorization, securing the control plane, and Pod security. We will also look at each of the security constructs with respect to their GKE implementations.

- **Hardening cluster security**: This section deep dives into GKE-specific security features that provide options for securing applications running inside the GKE cluster. This includes features such as private cluster, binary authorization, container-optimized OS, and more.

Technical requirements

There are two main technical requirements for this chapter:

- A valid **Google Cloud Platform** (**GCP**) account to go hands-on with GCP services: `https://cloud.google.com/free`

- Install Google Cloud SDK: `https://cloud.google.com/sdk/docs/quickstart`

Essential security patterns in Kubernetes

A Kubernetes cluster can run multiple types of workloads. This includes stateful applications, stateless applications, jobs, and DaemonSets. However, it is critical to secure these workloads from potential security attacks. Native Kubernetes provides some essential security constructs that focus on the fundamentals, including a request being sent to the cluster and how the request is authenticated and authorized. Additionally, it is important to understand how the master plane components are secured and how the pods running the applications can also be secured. We will cover these from a native Kubernetes standpoint, but their implementation in GKE will also be discussed. The first such security construct we will deep dive into is authentication.

Authentication

Authentication is the process of determining the identity of the user. It essentially confirms that the user is who they say they are and eventually provides access to eligible resources once authentication is successful.

Kubernetes supports two categories of authentication or user:

- User accounts
- Kubernetes service accounts

Let's look at these in more detail.

User accounts

By default, Kubernetes does not have any objects that can support normal user accounts. Hence, these can never be created through an API call. Normal or regular users in Kubernetes are created in any of the following ways:

- By an admin distributing private keys
- With a file that contains a list of usernames and their associated passwords
- Through external identity service providers

In the case of GKE, normal **user accounts** can be provisioned by Cloud IAM users. These user accounts are referred to as members. Members can also be defined as part of a G Suite domain or a Cloud Identity domain. It is also possible to add members or users to Cloud IAM by linking to an existing active directory through Google Cloud Directory Sync. In addition to Cloud IAM users, GCP service accounts are also considered members, like users. These are different from Kubernetes service accounts.

GCP service accounts are managed by Google Cloud IAM and specifically used if GCP resources need to have identities that are tied to an application or a virtual machine, instead of a human being. In contrast, Kubernetes service accounts provide an identity to a process running inside a pod and provides access to the cluster.

Kubernetes service accounts

Kubernetes service accounts are users that are managed by the Kubernetes API. This means that unlike regular user accounts, the service accounts can be created and managed through API calls. In fact, every namespace in Kubernetes has a default Kubernetes service account. These are automatically created by the API server. The *service account admission controller* associates the created service accounts with the running pods. In fact, service accounts are stored as secrets and are mounted onto pods when they're created. These secrets are used by processes running inside the pod for in-cluster access to the API server.

In addition, Kubernetes service accounts are used to create identities for long-running jobs where there is a need to talk to the Kubernetes API, such as running a Jenkins server. You can use the following CLI command to create a Kubernetes service account:

```
# Create kubernetes service account
kubectl create serviceaccount jenkins
```

The preceding command creates a `serviceaccount` object, generates a token for the service account, and creates a secret object to store the token. The secret bearing the token can be retrieved using the following CLI command:

```
# Get the definition of the service account
kubectl get serviceaccounts jenkins -o yaml
```

The preceding command will result in the following output:

```
apiVersion: v1
kind: ServiceAccount
metadata:
  # ...
secrets:
- name: jenkins-token-78abcd
```

The secret that's displayed under the secrets section will contain the public **Certificate Authority** (**CA** – an entity that issues digital certificates) of the API server, the specific namespace, and a signed **JSON Web Token** (**JWT**). The signed JWT can be used as the bearer account to authenticate the provided service account. This service account can eventually be used either for in-cluster communication or even to authenticate from outside the cluster, as in the case of a Jenkins server.

Every request to Kubernetes needs to be authenticated before it can serve requests. The incoming request is handled by the `kube-api` server by listening on port `443` using HTTPS. Authentication can be done in various ways. GKE supports the following authentication methods:

- OpenID Connect tokens
- x509 client certs
- Basic authentication using static passwords

OpenID Connect is a layer on top of the OAuth 2.0 protocol and allows clients to verify the identity of an end user by querying the authorization server. **x509 client certificates and static passwords** present a wider surface of attack than OpenID. In GKE, both x509 and static password authentication is disabled by default, specifically in clusters created with Kubernetes 1.12 and later. This helps improve the default security posture as the area of impact in the event of an attack is significantly reduced or lowered.

This completes this topic on authentication in Kubernetes. The next topic will cover authorization in Kubernetes.

Authorization

Authorization is the process of determining whether a user has permission to access a specific resource or perform a specific function. In Kubernetes, a user must be authenticated or logged in and authorized to access or use specific resources. It's generally recommended to enforce the principle of least privilege as a security best practice, as this ensures that a user only has the required level of access to the resource based on the access requirements.

Specific to GKE, a user authenticating via Cloud Identity can be authorized using two approaches. In fact, GKE recommends using both approaches to authorize access to a specific resource:

- Cloud **Identity and Access Management (IAM)**
- Kubernetes **Role-Based Access Control (RBAC)**

Cloud IAM is the access control system for managing GCP resources. Google Account, service account, and Google Group are entities that have an identity in Cloud IAM. Cloud IAM allows users to perform operations at the project level (such as listing all GKE clusters in the project) or at the cluster level (such as viewing the cluster) but specifically outside the cluster. This includes adding specific GKE security configuration options to an existing cluster or even to a new cluster. However, **Kubernetes RBAC** provides access to inside the cluster, even specifically at the namespace level. RBAC allows you to fine-tune rules to provide granular access to resources within the cluster.

To summarize, Cloud IAM defines who can view or change the configuration of a GKE cluster, while Kubernetes RBAC defines who can view or change Kubernetes objects inside the specific GKE cluster. GKE integrates Cloud IAM and Kubernetes RBAC to authorize users to perform actions on resources if they have the required permissions. Now, let's look at both authorization methods, starting with GKE authorization via Cloud IAM.

GKE authorization via Cloud IAM

There are three main elements that comprise Cloud IAM access controls. They are as follows:

- **Who**: This refers to authentication; specifically, the identity of the member making the request.

- **What**: This refers to authorization; specifically, the set of permissions that are required to authorize the request. Permissions cannot be directly assigned to members; instead, a set of permissions comprises a role that is assigned to members.

- **Which**: This refers to the resources that the request is authenticated and authorized to access. In the case of GKE, this refers to GKE resources such as the clusters or objects inside the cluster.

GKE provides several predefined Cloud IAM roles that provide granular access to Kubernetes engine resources. The following table summarizes the critical pre-defined IAM roles required to authorize or perform actions on GKE:

Role Name	Role Description
GKE Admin	Full access to Kubernetes API objects, along with the ability to perform actions specific to managing clusters.
GKE Cluster Admin	Provides access for managing clusters, such as creating, deleting, getting, listing, or updating clusters. No access to cluster resources or API objects.
GKE Cluster Viewer	Provides get and list access to clusters.
GKE Developer	Full access to Kubernetes API objects inside the Kubernetes cluster. Doesn't include operations that can be performed on the cluster directly, such as create, delete, and update.
GKE Host Service Agent User	Provides access to inspect firewall rules in the host project. Allows you to configure DNS resources, and also supports cluster management by allowing the Kubernetes Engine service account in the host project to configure shared network resources.
GKE Viewer	Read-only access to Kubernetes Engine resources.

You can always use custom roles with the minimum required set of permissions. This is specifically true in situations where the GKE pre-defined roles are too permissive or do not fit the use case at hand to meet the principle of least privilege.

Next, we will look at Kubernetes RBAC.

Kubernetes RBAC

Kubernetes RBAC is an authorization mechanism that can limit access to specific resources based on roles that have been assigned to individual users. RBAC is a native Kubernetes security feature that provides options to manage user account permissions. Kubernetes RBAC can be used as an added supplement to Cloud IAM. If Cloud IAM can define roles to operate on clusters and API objects within the cluster, then RBAC can be used to define granular access to specific API objects inside the cluster.

There are three main elements to Kubernetes RBAC. These are as follows:

- **Subjects**: This refers to a set of users or processes (including Kubernetes service accounts) that can make requests to the Kubernetes API.

- **Resources**: This refers to a set of Kubernetes API objects, such as Pod, Deployment, Service, and so on.

- **Verbs**: This refers to a set of operations that can be performed on resources such as get, list, create, watch, describe, and so on.

The preceding elements are connected by two RBAC API objects:

- **Roles**: Connects API resources and verbs
- **RoleBindings**: Connects Roles to subjects

Roles and RoleBindings can be applied at either the cluster level or at the namespace level. These will be discussed in the upcoming subsections, starting with *Roles*.

Roles

Roles connect API resources and verbs. There are two types of roles in RBAC. RBAC Roles are defined at the namespace level, while RBAC ClusterRole are defined at the cluster level. We'll look at these in the following sub-sections, starting with RBAC Roles.

RBAC Roles

The following is an RBAC Role that's been defined for a specific namespace:

```
apiVersion: rbac.authorization.k8s.io/v1
kind: Role
metadata:
  name: viewer
  namespace: production
rules:
apiGroups: [""]
  resources: ["pods"]
  verbs: ["get", "list"]
```

The definition represents a role of viewer that connects the resource pod with specific verbs, get and list, in the production namespace. Only one namespace can be defined per role. For core groups, the apiGroups section is optional. However, apiGroups should be specified for groups other than core groups. In addition, it is also possible to define a granular role where a specific resource name is also specified.

Multiple rules can be added to a role. Rules are additive in nature. An RBAC Role doesn't support deny rules. The following is an extension of the earlier RBAC Role, which now includes multiple rules and specifies a resource name:

```
apiVersion: rbac.authorization.k8s.io/v1
kind: Role
metadata:
  name: viewer
  namespace: production
rules:
apiGroups: [""]
  resources: ["pods"]
  verbs: ["get", "list"]
apiGroups: [""]
resources: ["ConfigMap"]
resourceNames: ["prodEnvironmentVariables"]
verbs: ["get", "list"]
```

In the preceding specification, the viewer RBAC Role can now perform `get` and `list` actions on *Pods* and `ConfigMap`. However, the operations on `ConfigMap` are strictly restricted to a specific `ConfigMap` named `prodEnvironmentVariables`.

This completes this sub-section on RBAC Role, one of the two possible RBAC roles. The other – *RBAC ClusterRole* – will be detailed in the following sub-section.

RBAC ClusterRole

RBAC ClusterRole grants permissions at the cluster level, so you don't need to define a specific namespace. The rest of the elements and their usage is the same as RBAC Role.

> **Namespace Scope versus Cluster Scope**
>
> There are specific resources that are scoped at the namespace level and others that are scoped at the cluster level. Pods, Deployments, Services, Secrets, ConfigMaps, PersistentVolumeClaim, Roles, and RoleBindings are namespace scoped. Nodes, PersistentVolume, CertificateSigningRequests, Namespaces, ClusterRoles, and ClusterRoleBindings are cluster scoped.

The following is the definition of an RBAC ClusterRole, where the intent is to define a role that can perform `list`, `get`, `create`, and `delete` operations against nodes in the cluster:

```
apiVersion: rbac.authorization.k8s.io/v1
kind: ClusterRole
metadata:
  name: node-administrator
rules:
apiGroups: [""]
    resources: ["nodes"]
    verbs: ["get", "list", "create", "delete"]
```

This completes this sub-section on Roles. The following sub-section explains how roles and users are tied through the *RoleBindings* Kubernetes API object.

RoleBindings

RoleBindings connect the subject to a role through a Kubernetes API object. There are two types of RoleBindings in RBAC. RBAC RoleBindings are defined at the namespace level, while RBAC ClusterRoleBindings are defined at the cluster level. Both will be discussed in the following sub-sections.

RBAC RoleBindings

The following is the definition of an RBAC RoleBinding that's been defined for a specific namespace that connects users to RBAC Roles:

```
apiVersion: rbac.authorization.k8s.io/v1
kind: RoleBinding
metadata:
  name: viewer-rolebinding
  namespace: production
subjects:
- kind: User
  name: joe@organization.com
  apiGroup: rbac.authorization.k8s.io
roleRef:
  kind: Role
  name: viewer
  apiGroup: rbac.authorization.k8s.io
```

The preceding RBAC RoleBinding has been defined for a production namespace and connects the elements defined under subjects to elements defined under `roleRef`. To be specific, the RBAC RoleBinding connects the user `joe@organization.com` to the `viewer` RBAC Role.

It's important to note that `kind` under the subject section can be of the *User*, *Group*, or *ServiceAccount* type. These values, from a GKE perspective, can either be from a Cloud IAM User, Cloud IAM service account, or a Kubernetes service account.

RBAC ClusterRoleBindings

RBAC ClusterRoleBindings bind subjects to RBAC ClusterRoles at the cluster level and are not restricted at the namespace level. You can only bind resources that are cluster scoped, not namespace scoped.

The following is the definition of RBAC `ClusterRoleBindings`, where the intent is to bind a specific admin user to the RBAC `ClusterRole`, called `node-administrator`, to perform operations against GKE nodes:

```
apiVersion: rbac.authorization.k8s.io/v1
kind: ClusterRoleBinding
metadata:
  name: node-administrator-clusterrolebinding
subjects:
- kind: User
  name: theadmin@organization.com
  apiGroup: rbac.authorization.k8s.io
roleRef:
  kind: ClusterRole
  name: node-administrator
  apiGroup: rbac.authorization.k8s.io
```

This completes this sub-section on RoleBindings, where both kinds of RoleBindings were explained. Overall, this also concludes the sub-section on Kubernetes RBAC and authorization in Kubernetes in particular. The upcoming sub-section discusses another key Kubernetes security construct – *control plane security* – which focuses on securing master control plane components.

Control plane security

As per GCP's shared responsibility model, GKE's **master control plane components** such as API server, `etcd` database, controller manager, and so on are all managed by Google. So, Google is responsible for securing the control plane, while the end user is responsible for securing nodes, containers, and pods.

Every GKE cluster has its own root CA. This CA represents an entity that issues a trusted certificate. This trusted certificate is used to secure the connection between machines. The root keys for a CA are managed by an internal service from Google. Communication between the master and the nodes in a cluster is secured based on the shared root of trust provided by the certificates issued by the CA. By default, GKE uses a separate per-cluster CA to provide certificates for the `etcd` databases within a cluster. Since separate CAs are used for each separate cluster, a compromised CA in one cluster cannot be used to compromise another cluster.

The Kubernetes API server and `kubelet` use secured network communication protocols such as TLS and SSH. They do this by using the certificates issued by the cluster root CA. When a new node is created in the cluster, the node is injected with a shared secret at the time of its creation. This secret is then used by its `kubelet` to submit certificate signing requests to the cluster root CA. This allows `kubelet` to get client certificates when the node is created, and new certificates when they need to be renewed or rotated. `kubelet` uses these client certificates to communicate securely with the API server.

You must periodically rotate the certificates or credentials to limit the impact of a breach. But sometimes, it might be difficult to strike a balance in terms of how often the credentials should be rotated. This is because the cluster API server will remain unavailable for a short period of time. Note that the credentials that are used by the API server and the clients can be rotated except for the `etcd` certificates, since these are managed by Google.

The following is the step-by-step process you should follow to rotate credentials:

1. The rotation process starts by creating a new IP address for the cluster master, along with its existing IP address.

2. `kube-apiserver` will not be available during the rotation process, but existing pods will continue to run.

3. New credentials are issued to the control plane as the result of a new IP address.

4. Once GKE has reconfigured the masters, the nodes are automatically updated by GKE to use the new IP and credentials. In addition, the node version is also updated to the closest supported version.

5. Each API client must be updated with the new address. Rotation must be completed for the cluster master to start serving with the new IP address and new credentials and to remove the old IP address and old credentials.

6. The master node will stop serving the old IP address.

7. If the rotation process is started but not completed within 7 days, then GKE will automatically complete the rotation.

Pods run on nodes and by default, pods can access the metadata of the nodes they are running on. This includes node secrets, which are used for node configuration. So, if a pod is compromised, the node secret also gets compromised, thus negatively impacting the entire cluster. The following steps should be taken to prevent such a compromised event and to protect cluster metadata:

- The service account tied to the nodes should not include the `compute.instance.get` permission. This blocks Compute Engine API calls to those nodes.

- The legacy Compute Engine API endpoint should be disabled (versions 0.1 and v1-beta-1) as these endpoints support metadata being queried directly.

- Use a workload identity to access Google Cloud services from applications running within GKE. This prevents pods from accessing the Compute Engine metadata server.

This completes this sub-section on how master control plane components are secured in GKE. Next, we'll look at how to secure pods running in a cluster by looking at pod security.

Pod security

One or more containers run inside a pod. By default, these containers can be deployed with privileged elevation. These are also known as privileged containers. **Privileged containers** have the root capabilities of a host machine and can access resources that can otherwise not be accessed by ordinary containers. The following are a few use cases for privileged containers:

- Running a Docker daemon inside a Docker container
- Requiring direct hardware access to the container
- Automating CI/CD tasks on an open source automation server, such as Jenkins

Running privileged containers is convenient but undesirable from a security perspective as it allows critical access to host resources. This privilege can be a disadvantage if it's exploited by cybercriminals. The attackers will have root access, which means they can identify and exploit software vulnerabilities and possible misconfigurations, such as containers with no authentication or minimum strength credentials. It is essentially a playground for coin miners to use this privilege for unauthorized needs.

There are two potential ways to define restrictions on what containers in a pod can do. They are as follows:

- Pod security context
- Pod security policy

Let's look at these options in more detail.

Pod Security Context

The security settings for a pod can be specified using the `securityContext` field in the pod specification. This applies to all the containers inside the pod and enforces the use of specific security measures. They can define whether privileged containers can run and whether the code in the container can be escalated to root privileges.

A security context can be defined both at the pod level and the container level. The container-level security context takes precedence over the pod-level security context. The following is an extract from the pod manifest YAML for `securityContext`:

```
apiVersion: v1
kind: Pod
metadata:
  name: my-pod
spec:
  securityContext:
    runAsUser: 3000
  containers:
  - name: nginx
    image: nginx
    securityContext:
      runAsUser: 1000
      allowPrivilegeEscalation: false
  - name: hello
    image: hello-world
```

The preceding specification represents a pod with two containers: `nginx` and `hello`. The `securityContext` definition on the pod specifies that processes inside containers run with a user ID of `3000`. It is important to specify a non-zero number as 0 in Linux as this represents a privileged user's user ID. Not specifying 0 takes away the root privilege of the code running inside the container. `securityContext` on the pod applies to all the containers inside the pod, unless each individual container has an optional `securityContext` defined. In that case, `securityContext` on the container takes precedence. So, in the preceding example, the `hello` container will run the process inside its container while using `3000` as the user ID, whereas the `nginx` container will run the process while using `1000` as the user ID.

Using allowPrivilegeEscalation on a Container

There are various ways we can use this field. One such scenario is that this field can be explicitly used where `securityContext` is not defined at the pod level but privilege escalation needs to be avoided at a specific container level.

Security contexts allow you to exercise control over the use of host namespaces, networking, filesystems, and volume types. A security context can be used to control additional security settings:

- **Provide specific capabilities**: If you don't want to give root capabilities, specific capabilities can also be specified at the container level. In the following example, `NET_ADMIN` allows you to perform network-related operations, such as modifying routing tables, enabling multicasting, and so on. `SYS_TIME` allows you to set the system clock:

```
. . .
spec:
  containers:
  - name: security-context-example
    image: gcr.io/demo/security-context-example
    securityContext:
      capabilities:
        add: ["NET_ADMIN", "SYS_TIME"]
```

- **Enable seccomp**: Blocks code that's running in containers from making system calls.

- **Enable AppArmor**: Restricts individual program actions using security profiles.

The downside of configuring `securityContext` for each pod and, sometimes, at each container level is that it involves a lot of effort, especially when hundreds of pods are involved in a cluster. This can be solved by using *Pod Security Policies*.

Pod Security Policies

A **Pod Security Policy** is a cluster-level resource that manages access for creating and updating pods, based on defined policies. A policy is a set of conditions that need to be met. A pod security policy makes it easier to define and manage security configurations separately. This allows you to apply security restrictions to multiple pods, without having to specify and manage those details in individual pod definitions.

Pod security Policies can enforce the following:

- **Disable privileged containers**: This can be disabled and can be optionally applied against a specific namespace and specific service account.

- **Enable read-only filesystems**: Containers such as web applications potentially write to a database and not necessarily to a filesystem. So, in such cases, `readOnlyRootFilesystem` can be set to `true`.

- **Enforce non-root users**: This can be enforced to not allow applications to run as root. You can do this by setting the `MustRunAsNonRoot` flag to `true`.

- **Prevent hostpath volumes**: This can be prevented by using `hostpath` for specific directories and not the entire filesystem.

There are two elements you need in order to define a Pod Security Policy:

- **PodSecurityPolicy object**: The `PodSecurityPolicy` object represents a set of restrictions, requirements, and defaults that are defined similar to a security context inside a pod. This object also specifies all the security conditions that need to be met for a pod to be admitted into a cluster. These rules are specifically applied when a pod is created or updated.

- **PodSecurityPolicy controller**: The `PodSecurityPolicy` controller is an admission controller. The admission controller validates and modifies requests against one or more Pod Security Policies. The controller essentially determines whether a pod can be created or modified.

Creating a PodSecurityPolicy object

If you need to create a `PodSecurityPolicy` object where privileged containers cannot be run in a specific namespace and by a specific service account, then you should follow these steps:

1. Define a `PodSecurityPolicy` Kubernetes object using the `pod-security-policy.yaml` file. The following is an example specification:

```
apiVersion: policy/v1beta1
kind: PodSecurityPolicy
metadata:
  name: my-pod-security-policy
spec:
  privileged: false #Prevents creation of privileged Pods
  runAsUser:
    rule: RunAsAny #Indicates any valid values to be used
```

Create the `PodSecurityPolicy` resource using the following CLI command:

```
# Create pod security policy
kubectl apply -f pod-security-policy.yaml
```

2. To authorize the specific `PodSecurityPolicy`, define a ClusterRole with the resource set to `podsecuritypolicies` and against the specific policy's resource name. An example `ClusterRole` specification for `my-cluster-role.yaml` is as follows:

```
apiVersion: rbac.authorization.k8s.io/v1
kind: ClusterRole
metadata:
  name: my-cluster-role
rules:
- apiGroups:
  - policy
  resources:
  - podsecuritypolicies
  verbs:
  - use
  resourceNames:
  - my-pod-security-policy
```

3. Create your `ClusterRole` using the following CLI command:

```
# Create ClusterRole
kubectl apply -f my-cluster-role.yaml
```

4. To authorize the created ClusterRole against a specific subject (which could be a service account) and, optionally, in a specific namespace, define a `RoleBinding`. An example specification for `my-role-binding.yaml` is as follows, where a RoleBinding is being applied to a specific service account:

```
# Bind the ClusterRole to the desired set of service
accounts
apiVersion: rbac.authorization.k8s.io/v1
kind: RoleBinding
metadata:
  name: my-role-binding
```

```
        namespace: my-namespace
    roleRef:
        apiGroup: rbac.authorization.k8s.io
        kind: ClusterRole
        name: my-cluster-role
    subjects:
        - kind: ServiceAccount
        name: sa@example.com
        namespace: my-namespace
```

5. Create your `RoleBinding` using the following CLI command:

```
# Create RoleBinding
kubectl apply -f my-role-binding.yaml
```

6. Enable the `PodSecurityPolicy` controller either at the time of cluster creation or while you're updating an existing cluster. The following are the CLI commands for both options:

```
# To enable at the time of cluster creation
gcloud beta container clusters create <cluster-name>
--enable-pod-security-policy


# To enable on an existing cluster
gcloud beta container clusters update <cluster-name>
--enable-pod-security-policy
```

7. If you ever need to disable the `PodSecurityPolicy` controller, use the following CLI command:

```
To disable PodSecurityPolicy controller
gcloud beta container clusters update <cluster-name>
--no-enable-pod-security-policy
```

In GKE, the `PodSecurityPolicy` controller is disabled by default or is not enabled at the time of cluster creation. So, it needs to be explicitly enabled. However, the controller should only be enabled once all the relevant `PodSecurityPolicy` objects have been defined, along with their authorization requirements. If there are multiple pod security policies, then these are evaluated alphabetically.

This concludes this section, which discussed the essential security concepts in the control plane (authentication), worker nodes, and deployments (authorization). This section also drew references to GKE and how these concepts are also implemented in GKE. The upcoming section focuses on specific GKE recommendations around hardening cluster security to ensure the applications running inside the cluster are secure. GKE offers certain features to support these recommendations, all of which will be outlined in detail.

Hardening cluster security in GKE

Securing the Kubernetes cluster should be your topmost priority when it comes to securing applications running inside your cluster. GKE supports many such features to harden the cluster. For example, the GKE control plane is patched and upgraded automatically as part of the shared responsibility model. In addition, node auto-upgrades are also enabled for a newly created GKE cluster.

The following are some key additional GKE features that can be used to secure and harden clusters. Some of these features are enabled by default while you're creating a GKE cluster:

- GKE supports a cluster type called **Private Cluster**, which provides options to restrict access to control planes and nodes. This needs to be specified at the time of cluster creation.

- GKE supports **container-optimized OS** images. It is a container-optimized OS that has been custom-built, optimized, and hardened specifically for running containers.

- GKE supports **shielded GKE nodes** as they help increase cluster security using verifiable node identities and integrity. This feature can be enabled on cluster creation or can be updated for an existing cluster.

- GKE allows you to enforce the use of **Network Policies** on a new or existing cluster. A network policy can restrict pod-to-pod communication within a cluster, thus reducing your footprint in the event of a security incident.

- GKE recommends using **binary authorization**, a process that ensures supply chain software security. Here, you have the option to exercise control so that only trusted images in the cluster are deployed.

- GKE can authenticate with other Google services and APIs through **Workload Identity**. This is the recommended way of doing things, instead of using the service account keys at the node level.

- GKE provides an additional layer of protection for sensitive data such as secrets by integrating with Google Secret Manager. **Secret Manager** is a GCP service that's used to secure API keys, passwords, certificates, and other sensitive data. GKE also supports the use of third-party secret managers such as HashiCorp Vault.

Each of the preceding GKE features will be covered in their respective sections. We will start with GKE private clusters.

GKE private clusters

GKE private clusters are one of the possible cluster configurations in GKE, especially when it comes to network isolation. This cluster configuration isolates node connectivity to the public internet. This includes both inbound traffic to the cluster and outbound traffic from the cluster. This is because the nodes inside the cluster will not have a public-facing IP address and will only have an internal IP address.

If nodes require outbound internet access, then a managed **Network Address Translation (NAT)** gateway can be used. Cloud NAT is GCP's managed NAT gateway. For inbound internet access, external clients can reach applications inside the cluster through services. The service type can either be of the `NodePort` or `LoadBalancer` type. If the service is of the `LoadBalancer` type, GCP's HTTP(S) load balancer can be used and will provide an external IP address to allow inbound traffic into the cluster. The key to GKE private clusters is the functionality of their control planes, since this is the main differentiating factor compared to a non-private cluster. We will look at this next.

Control plane in private clusters

In GKE, `kube-apiserver` is managed by the control plane. Google runs the control plane on a VM that is in a VPC network in a Google-owned project. In the case of a private cluster, the master control plane sitting on a Google-owned VPC network connects to your cluster's VPC network through VPC network peering. The traffic between the nodes and the control plane is routed through an internal IP address.

You can access the control plane through endpoints. In general, there are two types of endpoints:

- **Public endpoint**: This represents the external IP address of the control plane. Commands via the `kubectl` tool go through the public endpoint.

- **Private endpoint**: This represents the internal IP address in the control plane's VPC network. This is very specific to private clusters. The nodes in the private cluster communicate with the components in the control plane through internal IP addresses.

To summarize, a public cluster control plane has an internet-facing endpoint, while a private cluster control plane can be accessed both through private and public endpoints. In addition, a private cluster can only be created in a VPC-native mode (refer to *Chapter 8, Understanding GKE Essentials for Deploying Containerized Applications*). The level of access to a private cluster via endpoints can be controlled through one of the following three configurations:

- Public endpoint access disabled
- Public endpoint access enabled; authorized networks enabled for limited access
- Public endpoint access enabled; authorized networks disabled

Each of the preceding configurations will be discussed in detail in the upcoming sub-sections, starting with *Public endpoint access disabled*.

Public endpoint access disabled

This configuration represents a private GKE cluster with no access to a public endpoint. This is very secure as there is no access to the control plane via public internet. The cluster can only be accessed from the subnet and a secondary range used for pods. A VM in the same region can be added by updating the master authorized networks with the private IP of the VM in CIDR format.

If the cluster needs to be accessed from outside, then connect to the GKE private cluster's VPC network through Cloud VPN or Cloud Interconnect. The connection gets established through internal IP addresses. The list of internal IP addresses that can access the control plane can also be limited by using `master-authorized-networks`. This does not include public IP addresses as access to public endpoints is disabled.

Use the following CLI command if you need to create a private GKE cluster where you don't want client access to the public endpoint:

```
# For Standard Clusters
gcloud container clusters create my-private-cluster \
    --create-subnetwork name=my-subnet \
    --enable-master-authorized-networks \
    --enable-ip-alias \
    --enable-private-nodes \
    --enable-private-endpoint \
    --master-ipv4-cidr 172.20.4.32/28
```

The key flags in the preceding CLI command are as follows:

- `--enable-master-authorized-networks`: Access to the cluster control plane is restricted to the list of internal IP addresses. Cannot include external IP addresses.

- `--enable-private-nodes`: This indicates that the cluster nodes do not have external IP addresses.

- `--enable-private-endpoint`: This indicates that the cluster is only managed by the private IP address of the master API endpoint.

The next sub-section focuses on a configuration where public endpoint access is enabled but access is restricted.

Public endpoint access enabled; authorized networks enabled for limited access

This configuration represents a private GKE cluster configuration where there is restricted access to the control plane from both internal and external IP addresses. The specific set of internal and external IP addresses can be specified as part of the authorized networks. So, a machine with an external IP address can only communicate with a GKE Private Cluster if that IP address is included in the authorized networks.

Use the following CLI command if you need to create a private GKE cluster where there is limited access to a public endpoint:

```
# For Standard Clusters
gcloud container clusters create my-private-cluster-1 \
    --create-subnetwork name=my-subnet-1 \
    --enable-master-authorized-networks \
    --enable-ip-alias \
    --enable-private-nodes \
    --master-ipv4-cidr 172.20.8.0/28
```

Note that most of these flags are the same as they were in the previous sub-section, except for the omission of the `--enable-private-endpoint` flag. Omitting this flag implies that the cluster control plane can be reached both by private and public endpoints, but access is restricted only to the allowed IP address as part of the master authorized networks.

The next sub-section focuses on a configuration where public endpoint access is enabled and access is not restricted.

Public endpoint access enabled; authorized networks disabled

This is the default configuration option while creating a private GKE cluster. Essentially, the cluster will have access to the control plane from any IP address. This is the least restrictive option.

Use the following CLI command if you need to create a private GKE cluster where you wish there to be unrestricted access to the public endpoint:

```
# For Standard Clusters
gcloud container clusters create my-private-cluster-2 \
    --create-subnetwork name=my-subnet-2 \
    --no-enable-master-authorized-networks \
    --enable-ip-alias \
    --enable-private-nodes \
    --master-ipv4-cidr 172.20.10.32/28
```

Note that most of these flags are the same as the ones in the configuration where public endpoint access is enabled but master authorized networks are not enabled. As a result, there are no restricts in terms of the IP addresses that can access the control plane of the private GKE cluster either via a private endpoint or a public endpoint.

This completes this sub-section on private clusters, where nodes in the cluster can potentially be isolated or restricted from the public internet. The next topic shifts focus to container-optimized OS, which essentially protects the application by hardening the images that are used in containers with key security features. This feature is available in GKE.

Container-optimized OS

Container-optimized OS (also known as a `cos_containerd` image) is a Linux-based kernel that is custom-built from Google and is based on Chromium OS. It can continuously scan vulnerabilities at the kernel level or against any package of the OS. It can patch and update any package in case of a vulnerability. It is optimized and hardened specifically for running containers in production. The following are some of its key features:

- **Minimal OS footprint**: Doesn't include packages that are not required, thereby reducing the OS attack surface.

- **Immutable root system and verified boot**: The root filesystem is always mounted as read-only. This prevents attackers from making changes on the filesystem. Checksum is also computed at build time and verified by the kernel on each boot.

- **Support for stateless configuration**: The root filesystem can be customized to allow writes against a specific directory, such as `/etc/`. This is useful as you can allow write configuration at runtime, such as adding users to the filesystem. However, these changes are not persisted across reboots.

- **Security-hardened kernel**: Supports features such as seccomp and AppArmor to enforce fine-grained security policies.

- **Automatic updates**: Supports automatic updates for new security features or security patches for running GCE VMs.

- **Firewall**: By default, container-optimized OS doesn't allow any incoming traffic except SSH on port `22`.

Container-optimized OS ensures that the base image that's used to containerize the applications is secure and has a minimal footprint, but it is also important that these containers run on nodes that are equally secured or shielded. We will cover this in the next topic.

Shielded GKE nodes

Shielded GKE nodes is a GKE feature that increases cluster security by providing strong, verifiable node identity and integrity. These nodes are based on Compute Engine Shielded VMs.

> **Shielded VMs**
>
> Shielded VMs is a GCP feature where VM instances are ensured they won't be compromised at the boot or kernel level. GCP makes this possible by using secure boot and **virtual Trusted Platform Modules (vTPMs)**. Shielded VMs enforce and verify the digital signature of all the components at the time of boot process and halt the boot process on failure.

The shielded GKE nodes feature prevents the attacker from impersonating nodes in a cluster in the event of a pod vulnerability being exploited. If the shielded GKE nodes feature is enabled, the GKE control plane will cryptographically verify the following and limit the ability of the attacker to impersonate a node in the cluster:

- Every node in the GKE cluster is a GCE VM running in a Google data center.

- Every node is part of the cluster-provisioned managed instance group.

- `kubelet` authenticates with the node with a cluster-provisioned certificate.

You can use the following CLI commands to enable shielded GKE nodes in a new/existing cluster, to verify whether shielded GKE nodes are enabled, and to disable shielded GKE nodes:

```
# Enable Shielded GKE nodes on new cluster
gcloud container clusters create <cluster-name> --enable-
shielded-nodes


# Enable Shielded GKE nodes on existing cluster
gcloud container clusters update <cluster-name> --enable-
shielded-nodes


# Verify that Shielded GKE nodes are enabled (check for enabled
under shieldedNodes as true)
gcloud container clusters describe <cluster-name>


# Disable Shielded GKE nodes (This will recreate the control
plane and nodes thus leading to downtime)
gcloud container clusters update <cluster-name> --no-enable-
shielded-nodes
```

There is no extra cost in running shielded GKE nodes. However, they produce more logs than regular nodes, thus leading to an overall increase in costs with respect to Cloud Logging. The next topic explains another GKE security feature where the surface area of the attack is reduced in case of a security threat, by restricting the traffic among pods in a cluster.

Network Policies – restricting traffic among pods

All the pods within a Kubernetes cluster can communicate with each other. However, Kubernetes provides an option for when the traffic between pods needs to be controlled at the IP address or port level. This thought process is strongly recommended to ensure that the entire cluster is not compromised, and that the surface area is controlled in case of a security attack. Kubernetes Network Policies helps you restrict traffic among pods within the cluster.

Network Policies in Kubernetes allow you to specify how a pod can communicate with various network entities based on pods with matching label selectors, namespaces with matching label selectors, or specific IP addresses with port combinations (including the ability to specify exception IP addresses). This can be defined for either ingress or egress traffic flowing in both directions.

GKE provides options to enforce the use of a network policy when a cluster is created, like so:

```
# Enforce network policy for a new GKE cluster
gcloud container clusters create <cluster-name> --enable-
network-policy
```

Optionally, we can enforce this on an existing cluster by using the following CLI commands:

```
# Enable add-on to enforce network policy on existing cluster
gcloud container clusters update <cluster-name> --update-
addons=NetworkPolicy=ENABLED


# Enforce network policy after enabling add-on for existing
cluster. This will recreate the cluster node pools
gcloud container clusters update <cluster-name> --enable-
network-policy
```

A sample network policy can be found at `https://kubernetes.io/docs/concepts/services-networking/network-policies/`.

In addition to specifying a pinpointed policy, Kubernetes allows you to define default network policies. The following are some of the supported default policies:

- Default deny all ingress traffic
- Default deny all egress traffic
- Default deny all ingress and all egress traffic
- Default allow all ingress traffic
- Default allow all egress traffic

If a specific network policy or a default policy is not defined, then the cluster will allow both ingress and egress traffic to and from pods.

The next topic details another key GKE feature known as Binary Authorization, which can exercise control to ensure only trusted images are deployed to the GKE cluster.

Binary Authorization

Binary Authorization is a deploy-time security service provided by Google. It ensures that only trusted containers are deployed in the GKE cluster using deployment policies. The goal of the policy is to determine which images to allow and which to exempt. To accomplish this goal, Binary Authorization integrates with Container Analysis – a GCP service that scans container images stored in a Container Registry for vulnerabilities. In addition, **Container Analysis** also stores trusted metadata that's used in the authorization process.

Binary Authorization policies are security-oriented and comprise one or more rules. Rules are constraints that need to pass before the images can be deployed to the GKE cluster. An attested image is one that has been verified or guaranteed by an *attestor*. The most common rule that is used is the need for a digitally signed attestation to verify whether the image has been attested. When a container image is built through Cloud Build, the image's digest is digitally signed by a signer, which creates an attestation. At the time of deployment, Binary Authorization enforces the use of an attestor to verify the attestation. Binary Authorization only allows *attested* images to be deployed to the cluster. Any unauthorized images that do not match the Binary Authorization policy are rejected. Additionally, a *Denied by Attestor* error can also be returned if no attestations are found that were valid and were signed by a key trusted by the attestor. To overcome the *Denied by Attestor* error, create an attestation and submit it to Binary Authorization.

The following are some common use cases that include attestations:

- **Build verification**: To verify whether the container image was built by a specific build system or from a specific **continuous integration** (**CI**) pipeline.
- **Vulnerability scanning**: To verify whether the CI-built container image has been scanned for vulnerabilities by Container Analysis and the findings have been defined at an acceptable level.

Configuring Binary Authorization is a multi-step process. The following is a high-level summary of the steps involved:

1. Enabled the required APIs. This includes APIs for GKE, Container Analysis, and Binary Authorization.
2. Create a GKE cluster with binary authorization enabled.
3. Set up a note. This is a piece of metadata in Container Analysis storage that is associated with an attestor.

4. Set up cryptographic keys using PKIX keys, to securely verify the identity of attestors; only enforce verified parties to authorize a container image. **Public-Key Infrastructure (PKIX)** keys refer to public key certificates defined in the X.509 standard.

5. Create an attestor; that is, a person or process that attests to the authenticity of the image.

6. Create a Binary Authorization policy. The default policy is to allow all images. The other option includes denying all images or denying images from a specific attestor.

7. Optionally, images can be configured so that they're exempt from the binary authorization policy.

As we mentioned previously, Binary Authorization can deny images from being deployed if the policy conditions are violated or not met. However, you can specify the break-glass flag as an annotation in the pod deployment, which allows pod creation even if the images violate the policy. The break-glass annotation flag is also logged and can be identified by incident response teams through audit logs while they're reviewing or debugging the deployments. The following is a snippet of a pod specification that includes the break-glass flag annotation:

```
apiVersion: v1
kind: Pod
metadata:
  name: my-break-glass-pod
  annotations:
    alpha.image-policy.k8s.io/break-glass: "true"
```

This concludes this topic on Binary Authorization. The next topic details another key GKE security feature that allows Google Cloud IAM service accounts to be used as Kubernetes service accounts through Workload Identity, thus providing more secure access to GCP services from applications running inside the GKE cluster.

Workload Identity

GKE clusters can run applications that might need access to Google-specific APIs, such as compute APIs, storage APIs, database APIs, and more. GKE recommends using *Workload Identity* to access GCP services from applications running within GKE. Workload Identity allows you to use a Kubernetes service account as a Google service account. This allows each application to have distinct identities and fine-grained authorization.

Workload Identity uses the concept of a cluster workload identity pool, which allows Google Cloud IAM to trust and understand Kubernetes service account credentials. The cluster's workload identity pool will be set to `PROJECT_ID.svc.id.goog` and is automatically created at the project level. In such a scenario, Cloud IAM will authenticate the Kubernetes service account with the following member name:

```
serviceAccount:PROJECT_ID.svc.id.goog[K8S_NAMESPACE/KSA_NAME]
# PROJECT_ID.svc.id.good - workload identity pool on the
cluster
# KSA_NAME Kubernetes - service account making the request
# K8S_NAMESPACE Kubernetes - namespace with Kube SA is defined
```

The preceding member's name is unique due to the cluster's Workload Identity pool, service account name, and Kubernetes namespace. So, multiple service accounts with the matching three tuples will map to the same member name.

Enabling Workload Identity

Follow these steps to enable Workload Identity on a GKE cluster:

1. Navigate to **APIs & Services** under the GCP console. Then, search for the `IAM service account Credentials` API and enable it.

2. Create a new cluster with Workload Identity enabled via the following CLI command:

    ```
    # Create cluster with workload identity enabled
    gcloud container clusters create <CLUSTER_NAME> \
      --workload-pool=<PROJECT_ID>.svc.id.goog
    ```

3. Update an existing cluster with Workload Identity enabled via the following CLI command:

    ```
    # Update existing cluster with workload identity enabled
    gcloud container clusters update <CLUSTER_NAME> \
      --workload-pool=<PROJECT_ID>.svc.id.goog
    ```

This concludes this section on Workload Identity, as well as this major section on the key GKE security features that are recommended by Google to harden cluster security.

Summary

In this chapter, we discussed some fundamental security concepts from a native Kubernetes or K8s standpoint. Each of these concepts was extended as we looked at their equivalent usage or implementation in GKE. Later, we did a deep dive into certain GKE-specific security features that are critical to hardening cluster security. This included using node auto upgrades to ensure that the nodes are running the latest version of Kubernetes, or using Google's container-optimized OS instead of a general-purpose Linux distribution system. We also looked at using private clusters, where access to the cluster master can be restricted for enhanced security or can be controlled so that it's only accessed from authorized networks. We also looked at Binary Authorization, which ensures that only trusted images can be deployed to the cluster, and Workload Identity, which allows us to use a Cloud IAM service account as a Kubernetes service account, thus providing more flexibility in terms of which applications in the GKE cluster can easily interact with other GCP services, such as Cloud Storage, Secret Management, and more.

In the next chapter, we'll look at services that are tied to Cloud Operations, along with a specific feature that was introduced in Google Cloud to track the reliability of services: Service Monitoring. This specific feature/option links the SRE technical practices (SLIs, SLOs, and error budget) to the features that are available in Google Cloud Operations so that we can monitor services and alert others about their reliability.

Points to remember

The following are some important points to remember:

- GCP service accounts are used if GCP resources must have an identity that is tied to an application or a virtual machine.

- Kubernetes service accounts are users that are managed by the Kubernetes API.

- Cloud IAM defines who can view or change the configuration of a GKE cluster and Kubernetes RBAC defines who can view or change Kubernetes objects inside the specific GKE cluster.

- Workload Identity is used to access Google Cloud services from applications running within GKE. This prevents pods from accessing the Compute Engine metadata server.

- In RBAC, a Role connects API resources and verbs. An RBAC Role is cluster-wide scoped, while an RBAC ClusterRole is namespace scoped.

- In RBAC, RoleBindings connect Roles to subjects. A RoleBinding is cluster-wide scoped, while a ClusterRoleBinding is namespace scoped.

- Every GKE cluster has its own root **CA**.

- Pod Security Context and Pod Security Policy are two ways we can define restrictions regarding what the containers inside a pod can do.

- A GKE Private Cluster allows you to restrict access to control planes and nodes.

- The break-glass flag is used in deployments as an annotation; it allows pod creation, even if the images violate a policy.

- `enable-private-nodes`: The nodes do not have external IP addresses.

- `enable-private-endpoint`: The cluster is managed by the private IP address of the master endpoint.

- `enable-master-authorized-networks`: Access to the cluster's public endpoint is restricted to a specific set of source IP addresses.

- Container Analysis is a service that provides vulnerability scanning and metadata storage for software artifacts.

- Container Analysis stores trusted metadata that's used in the authorization process.

- Binary Authorization allows or blocks images from being deployed to GKE based on a policy you've configured.

- A container-optimized OS or `cos_containerd` image is a Linux-based kernel that can continuously scan for vulnerabilities at the kernel level.

- Shielded GKE nodes increase cluster security by using verifiable node identity and integrality. This can be enabled using the `--enable-shielded-nodes` option.

- You can restrict traffic among pods with Network Policies.

- You can configure a secret manager that has been integrated with GKE clusters.

- You can use admission controllers to enforce a Pod Security Policy.

- In terms of Workload Identity, you can use K8's service account and namespace as a GCP service account to authenticate GCP APIs.

Further reading

For more information on GCP's approach to DevOps, read the following articles:

- **Kubernetes**: `https://kubernetes.io/docs/home/`

- **Google Kubernetes Engine**: `https://cloud.google.com/kubernetes-engine`

Practice test

Answer the following questions:

1. Network Policies are used to restrict traffic among which of the following?

    a) Deployments

    b) Containers

    c) Pods

    d) Container images

2. Select the RBAC option that connects a user and a role:

    a) UserRoleBinding

    b) RoleBindings

    c) Roles

    d) RoleUserBinding

3. In a private cluster, which Google service can download a Docker image?

    a) Cloud Build

    b) Cloud Source Repository

    c) Elastic Container Registry

    d) Container Registry

4. What will happen if the process of rotating credentials started but never completed?

    a) GKE will not complete the cluster rotation.

    b) GKE will pause the cluster rotation.

    c) GKE will complete the cluster rotation in 7 days.

    d) GKE will instantly complete the cluster rotation.

5. Which of the following possible policies will disable privileged containers?

    a) Network Policy

    b) Pod Security Policy

    c) Network Security Policy

    d) Pod Policy

6. Select the GKE Role that allows you to manage clusters, including creating, deleting, getting, listing, or updating clusters. No access is given to cluster resources or API objects:

   a) GKE Admin

   b) GKE Cluster Admin

   c) GKE Developer

   d) GKE Cluster Developer

7. With regards to a **Pod Security Policy** (**PSP**), select the order of operations:

   a) Enable PSP Controller, Create PSP, Define Authorization Requirements

   b) Create PSP, Enable PSP Controller, Define Authorization Requirements

   c) Create PSP, Define Authorization Requirements, Enable PSP Controller

   d) Enable PSP Controller, Define Authorization Requirements, Create PSP

8. If a specific network policy or a default policy is not defined, then which of the following is true?

   a) Deny all ingress and all egress traffic.

   b) Allow all ingress and all egress traffic.

   c) Deny all ingress and allow all egress traffic.

   d) Allow all ingress and deny all egress traffic.

9. Select the option that enforces a deploy time policy to GKE:

   a) Cloud IAM Policies

   b) AppArmor

   c) Cloud Armor

   d) Binary Authorization

10. The *service account admission controller* associates the created service accounts with the running pods. How are the service accounts stored and accessed?

    a) Stored as plain text and accessed as environment variables at runtime

    b) Stored as a Kubernetes secret and accessed through the key management service

    c) Stored as a Kubernetes secret and accessed as an environment variable at runtime

    d) Stored as plain text and accessed through the key management service

Answers

1. (c): Pods

2. (b): RoleBindings

3. (d): Container Registry

4. (c): GKE will complete the cluster rotation in 7 days.

5. (b): Pod Security Policy

6. (b): GKE Cluster Admin

7. (c): Create PSP, Define Authorization Requirements, Enable PSP Controller

8. (b): Allow all ingress and all egress traffic.

9. (d): Binary Authorization

10. (c): Stored as a Kubernetes secret and accessed as an environment variable at runtime

10
Exploring GCP Cloud Operations

Reliability is the most critical feature of a service or a system. **Site Reliability Engineering (SRE)** prescribes specific technical tools or practices that help in measuring characteristics that define and track reliability, such as **SLAs, SLOs, SLIs**, and **error budgets**.

In *Chapter 2, SRE Technical Practices – Deep Dive*, we discussed the key constructs of SLAs in detail, the need for SLOs to achieve SLAs, the guidelines for setting SLOs, and the need for SLIs to achieve SLOs. In addition, we learned about the different types of SLIs based on user journey categorization, different sources to measure SLIs, the importance of error budgets, and the ways to set error budgets to make a service reliable.

This raises a series of fundamental questions:

- How can we observe SLIs for a service so that the SLOs are not violated?

- How can we track whether our error budgets are getting exhausted?

- How can we maintain harmony between the key SRE technical tools?

SRE's answer to the preceding questions is observability. Observability on **Google Cloud Platform** (**GCP**) is established through operations. From Google's point of view, Cloud Operations is about monitoring, troubleshooting, and improving application performance in the Google Cloud environment. The key objectives of Cloud Operations are as follows:

- Gather logs, metrics, and traces from any source.

- Query the captured metrics and analyze traces.

- Visualize information on built-in or customizable dashboards.

- Establish performance and reliability indicators.

- Trigger alerts and report errors in situations where reliability indicators are not met, or issues are encountered.

Google achieves key objectives of operations through a collection of services called Cloud Operations. Cloud Operations is a suite of GCP services that includes Cloud Monitoring, Cloud Logging, Error Reporting, and **Application Performance Management** (**APM**). Furthermore, APM includes Cloud Debugger, Cloud Trace, and Cloud Profiler. This chapter will explore services tied to Cloud Operations. Post that, we will focus on a specific feature that was introduced in Google Cloud to track the reliability of services through Service Monitoring. This specific feature/option links the SRE technical practices (SLIs, SLOs, and Error Budget) to features available in Google Cloud Operations that monitor the service and tell us about its reliability.

In this chapter, we're going to cover the following main topics:

- **Cloud Monitoring**: Workspaces, dashboards, Metrics explorer, uptime checks, and alerting.

- **Cloud Logging**: Audit Logs, Logs Ingestion, Logs Explorer, and Logs-Based Metrics.

- **Cloud Debugger**: Setting up, Usage, Debug Logpoints, and Debug Snapshots.

- **Cloud Trace**: Trace Overview, Trace List, and Analysis Reports.

- **Cloud Profiler**: Profile Types.

- **Binding SRE and Cloud Operations**: We will measure service reliability using Cloud Operations by linking them to SRE technical practices via a hands-on lab.

Cloud Monitoring

Cloud Monitoring is a GCP service that collects metrics, events, and metadata from multi-cloud and hybrid infrastructures in real time. Cloud Monitoring helps us understand how well our resources are performing and if there is something wrong that requires immediate attention. Cloud Monitoring is a medium through which SRE best practices can be implemented and to ensure that applications are meeting their set SLAs. Cloud Monitoring consists of out-of-the-box dashboards. These can be used to visualize insights into key factors that impact SLIs and SLOs such as latency, throughput, and more. Cloud Monitoring is also critical for incident management as alerts can be generated from key metrics, and these alerts can be sent as notifications to configured notification channels.

This section on Cloud Monitoring deep dives into several key areas/properties, such as workspaces, dashboards, Metrics explorer, uptime checks, alerting, access controls, and the Monitoring agent. We will start with workspaces.

Workspaces

Google Workspace is a centralized hub that's used to organize and display monitoring information about GCP and AWS resources. Workspace provides a centralized view and acts as a single point of entry to resource dashboards, uptime checks, groups, incidents, events, and charts. Google describes this centralized view as a *single pane of glass* (please refer to the following screenshot). The actions that can be performed against a workspace include the ability to view content, which is controlled by **Identity and Access Management (IAM)**:

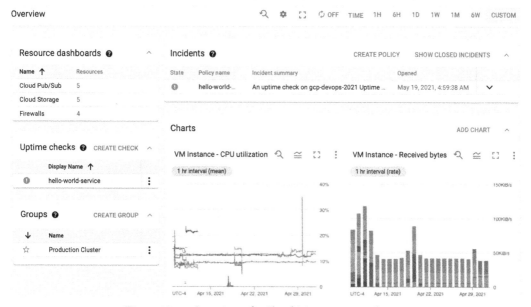

Figure 10.1 – Overview of a Cloud Monitoring workspace

The following is some key terminology we need to know about before we elaborate on the relationship between workspaces and projects:

- **Host project**: This refers to the project where Workspace is created.

- **Monitored project**: This refers to the GCP projects or AWS accounts that the workspace can monitor.

- **AWS connector project**: This refers to the GCP project that connects the monitored AWS account to the workspace.

The upcoming subsection provides insights into the relationship between Workspace and Project.

Workspace/project relationship

The following are some key pointers with respect to a workspace/project relationship:

- A workspace is always created inside a project. This is known as the host project.

- A workspace is part of a single host project and is named after the host project.

- A workspace can monitor resources from multiple monitored projects simultaneously. This could include about 200 GCP projects/AWS accounts.

- A workspace can access other monitored projects' metric data, but the actual data lives in monitored projects.

- A monitored project can only be associated with a single workspace, and a monitored project can be moved from one workspace to another.

- Multiple workspaces can be merged into a single workspace.

- There is no charge associated with creating a workspace. However, charges with respect to logging and ingesting metric data is charged to the billing account associated with the *monitored projects*.

> **Tip – how to connect an AWS account to a workspace**
>
> A GCP connector project is required. This can be an existing project, or an empty project (preferred) created for this purpose. GCP connector project needs to be under the same parent organization as the workspace. A billing account should be tied to the connector project and this account will be charged to monitor the resources under the AWS account.

The following section discusses potential strategies for creating a workspace.

Workspace creation – strategies

In a real-time scenario, it is possible to have multiple projects where the projects are either differentiated by customers or differentiated by environment types such as development, test, staging, and production. Given that a workspace can monitor resources from multiple projects, the strategy/approach that's taken to create the workspace becomes critical. There are multiple strategies we can follow to create a workspace, as detailed in the following sections.

A single workspace for all monitored projects

Information about all monitored project resources is available from within a single workspace. The following diagram represents a single workspace that monitors an application, app, that's been deployed across multiple projects, such as app-dev, app-test, and app-prod. These have been categorized by environment type. This approach has its own pros and cons.

The pro is that the workspace provides a single pane of glass for all the resources tied to the application across multiple projects representing multiple environment types. The con is that a non-production user can access resources from a production project, and this might not be acceptable in most cases. This approach is not suitable for organizations that have strict isolation between production and non-production environments:

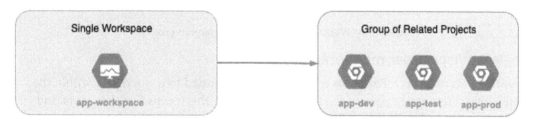

Figure 10.2 – A single workspace for all related projects

Workspace per group of monitored projects

A workspace will monitor a specific group of projects. There could be more than one workspace monitoring the available projects. The following diagram is an alternative workspace creation strategy compared to the one shown in the preceding diagram. Specifically, the following diagram represents two workspaces where one workspace monitors the non-production projects and the second workspace monitors the production project. Access controls to that specific group can be controlled by the host project of the individual workspace. This allows us to differentiate between users across environment types, such as production versus non-production or even across customers:

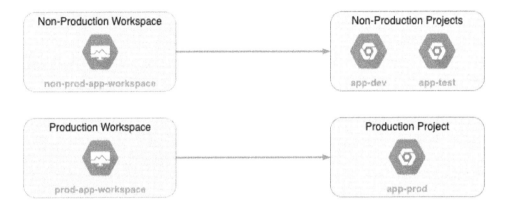

Figure 10.3 – Workspace per group of monitored projects

Single workspace per monitored project

Essentially, every project that needs to be monitored is hosted by a workspace within the same project. This can be seen in the following diagram. This means that the source and the host project will be the same. This approach provides the most granular control in terms of providing monitoring access to the project resources. However, this might also only provide a slice of information if an application is spread across multiple projects:

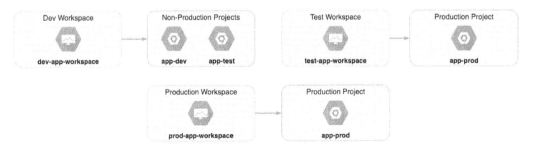

Figure 10.4 – Single workspace per monitored project

Important note – Workspace management

One or more GCP project(s) or AWS account(s) can either be added or removed from the workspace. In addition, all projects within a selected workspace can be merged into the current workspace, but the configuration will be deleted in the selected workspace. These actions are performed through the settings page of Cloud Monitoring. A workspace can only be deleted if the workspace host project is deleted.

This concludes this subsection on workspace creation strategies. The preferred strategy depends on the organizational need. Workspace basic operations include creating a workspace, adding project(s) to a workspace, moving projects between workspaces, and merging workspaces. Detailed instructions on how to create a workspace can be found at `https://cloud.google.com/monitoring/workspaces/create`. The upcoming subsection discusses IAM roles, which are used to determine who has access to monitor the resources inside the monitoring workspace.

Workspace IAM roles

The following are the IAM roles that can be applied to a workspace project so that we can view monitoring data or perform actions on the workspace:

- **Monitoring Viewer**: Read-only access to view metrics inside a workspace.

- **Monitoring Editor**: Monitoring Viewer, plus the ability to edit a workspace, create alerts, and have write access to Monitoring Console and Monitoring API.

- **Monitoring Admin**: Monitoring Editor, plus the ability to manage IAM roles for the workspace.

- **Monitoring Metric Writer**: A service account role that's given to applications instead of humans. This allows an application to write data to a workspace but does not provide read access.

This concludes our quick insight into workspaces and their concepts, such as workspace creation strategies. Next, we will look at dashboards.

Dashboards

Dashboards provide a graphical representation of key signal data, called metrics, in a manner that is suitable for end users or the operations team. It's recommended that a single dashboard displays metrics depicting a specific viewpoint (for example, serverless resources with a specific label) or for a specific resource (for example, persistent disks, snapshots, and so on). There are two types of dashboards: predefined dashboards and custom dashboards.

The following screenshot is of the **Dashboards Overview** page in Cloud Monitoring. This page displays the list of available dashboards, categorized by dashboard types, and provides quick links to the most recently used dashboards:

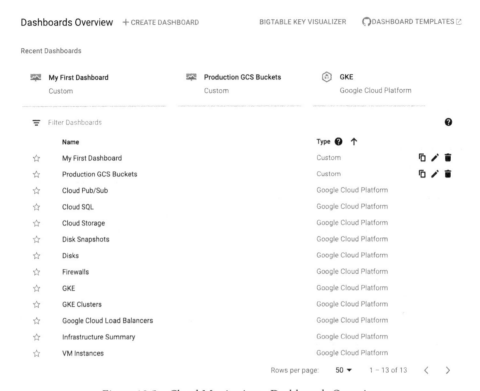

Figure 10.5 – Cloud Monitoring – Dashboards Overview

Cloud Monitoring supports both predefined dashboards and custom dashboards. The upcoming subsection provides an overview of the different types of dashboards and steps for how to create a custom dashboard.

Predefined dashboards

Cloud Monitoring provides a set of predefined dashboards that are grouped by a specific GCP resource such as firewalls or GKE Clusters. These dashboards are categorized by **Type**. This is set to **Google Cloud Platform**, which is maintained by Google Cloud. They do not require any explicit setup or effort to configure.

However, predefined dashboards are not customizable. These dashboards are organized in a specific manner with a set of predefined filters from the context of the dashboard. Users cannot change the contents of the view or add a new filter criterion. Users can only use the predefined filters to control the data being displayed.

Custom dashboards

Users or operation teams can create a **custom dashboard** that displays specific content of interest. These dashboards are categorized by **Type** set to **Custom**. Content is added by configuring one or more widgets. There are multiple types of widgets. Dashboards can either be created from Google Cloud Console or via the Cloud Monitoring API. In addition, the Cloud Monitoring API allows you to import a dashboard configuration from GitHub and modify it as needed.

Custom dashboards represent information about a metric using a chart. This chart displays raw signal information from a metric that's aligned across a configurable time window. Each chart is of a specific widget type. Cloud Monitoring supports multiple widget types such as Line, Stacked area, Stacked bar, Heatmap, Gauge, Scorecard, and Textboxes. Let's take a brief look at the different types of widgets:

- Line charts, stacked area charts, and stacked bar charts are best utilized to display time series data. Each of these widget types can be configured so that they're displayed in Color/X-Ray/Stats/Outlier mode, along with an optional legend, using the display view options.

- Heatmap charts are used to represent metrics with a distribution value.

- Gauges display the most recent measurement in terms of a number. This is represented by a thick line around the gauge. This is visually categorized across good, warning, and danger zones.

- Scorecards are similar to gauges as they display the recent measurement in terms of a number, but they can be visually depicted using a different view other than a gauge, such as a spark line, spark bar, icon, or value.

- Textboxes allow us to add any custom information, such as quick notes or links, concerning the relevant resources.

The upcoming subsection will show you how to create a custom dashboard.

Creating a custom dashboard

Follow these steps to create a custom dashboard from the GCP console:

1. Navigate to **Cloud Monitoring | Dashboards** and select the **Create Dashboard** option. Name the dashboard VM Instance - Mean CPU Utilization.

2. Select a chart type or widget. This will open **Metrics explorer** on the left-hand pane and will add the chart to the dashboard.

3. Select the options to choose a resource type, metric type, and grouping criteria. Then, **Save** the dashboard to add the chart type.

4. Redo the preceding steps to add multiple charts to the same dashboard.

The following screenshot shows a custom dashboard depicting the mean CPU utilization for all the VM instances, along with seven possible widget types:

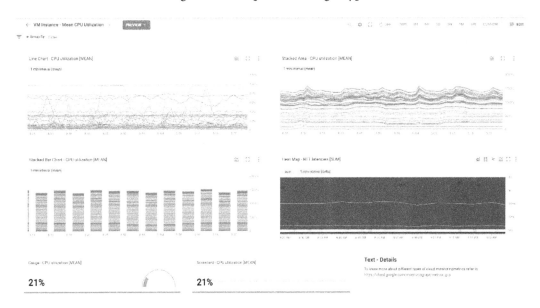

Figure 10.6 – Custom dashboard with seven possible widget types

This concludes this section on Cloud Monitoring dashboards. The next section focuses on using *Metrics explorer* as an option to explore predefined and user-created metrics.

Metrics explorer

Metrics is one of the critical sources for monitoring data. Metrics represents numerical measurements of resource usage or behavior that can be observed and collected across the system over many data points at regular time intervals. There are about 1,000 pre-created metrics in GCP. This includes CPU utilization, network traffic, and more. However, some granular metrics such as memory usage can be collected using an optional Monitoring agent. Additionally, custom metrics can be created either through the built-in Monitoring API or through OpenCensus – an open source library used to create metrics. It is always recommended to check if a default or pre-created metric exists before creating a custom metric.

Metrics explorer provides options for exploring existing metrics (either predefined or user-created), using metrics to build charts, adding charts to an existing or new dashboard, sharing charts via a URL, or retrieving chart configuration data in JSON format. Metrics explorer is an interface that provides a DIY approach to building charts as you can select a metric of your choice.

The following screenshot shows the **Metrics explorer** section, which charts the **CPU Utilization** metric for a **VM Instance**, grouped by system state. The left-hand side displays the configuration region, while the right-hand side depicts the chart for the selected metric. The configuration region has two tabs:

- **Metric**: This tab is used to select the metric and explore it.

- **View Options**: This tab is used to change the chart's display characteristics:

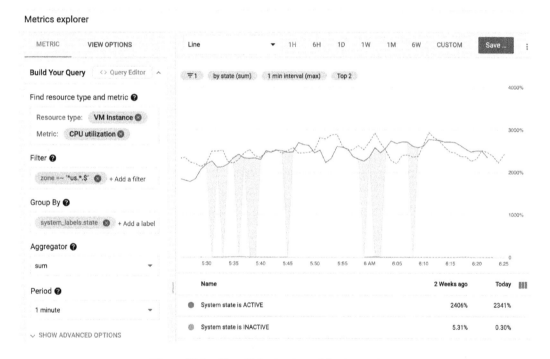

Figure 10.7 – Cloud Monitoring – Metrics explorer

The following section discusses the available options for configuring a metric using Metrics explorer.

Understanding metric configuration via Metrics explorer

To configure a metric for a monitored resource, we can use the following options.

Resource type and **Metric** option can be selected in either of the following ways:

- **Standard Mode**: Select a specific metric type or browse the available metric types based on a specific GCP resource.

- **Direct Filter Mode**: Manually enter a metric type and resource type in a text box.

The **Filter** option can be used to filter out the results based on the filter criterion. The filter criterion can be defined using the available operators or regular expressions. There are two possible filter types:

- **By Resource Label**: The available filter fields are specific to the selected resource type (for example, the VM Instance resource type will have `project_id`, `instance_id`, and `zone` as the available filter options).

- **By Metric Label**: Refers to project-wide user-created labels.

The **Group By** option can be used to group time series data by resource type and metric label. This creates new time series data based on the combination of group by values.

The **Aggregator** option can be used to describe how to aggregate data points across multiple time series. Common options include min, max, sum, count, and standard deviation. By default, the aggregation results in a single line by applying the aggregator across all the time series. If **Group By** labels are selected, the aggregation results in a time series for each combination of matching labels.

The **Period** option can be used to determine the time interval for which aggregation takes place. The default selection is 1 minute.

The **Aligner** option can be used to bring the data points in each individual time series into equal periods of time.

Additional options include the following:

- **Secondary Aggregator**: Used in charts with multiple metrics

- **Legend Template**: For better readability

Multiple **View Options** can be used to plot metrics, and these are distinguished by the available chart modes, which are as follows:

- **Color mode**: This is the default mode where graph lines are shown in color.

- **X-Ray mode**: Shows graph lines in a translucent gray color but with brightness in the case of overlapping bands.

- **Stats mode**: Shows common statistical values such as the 5th percentile, 95th percentile, average, median, and more.

- **Outlier mode**: Allows you to choose a number of time series to display, along with the option to rank time series by ordering them from the top or bottom.

Additionally, each chart mode supports the ability to specify a specific threshold and allows you to compare past time series data. In addition, it is possible to apply a log scale to the y axis for better separation between larger values in datasets where some values are much larger than the others.

> **Tip – Monitoring Query Language (MQL) – Advanced option to create charts**
>
> Cloud Monitoring supports **MQL**, an advanced option for creating a chart with a text-based interface and an expressive query language that can execute complex queries against time series data. Potential use cases include the ability to select a random sample of time series or compute the ratio of requests, resulting in a particular class of response codes.

This completes this section on Metrics explorer, which allows users to explore predefined and custom metrics. These can potentially be used to create charts. The options related to configuring a metric were also discussed in detail. The upcoming section focuses on uptime checks – an option for validating whether a service is functioning.

Uptime checks

Uptime check is a Cloud Monitoring feature where periodic requests are sent to monitor a resource to check if the resource is indeed up. Uptime checks can check the uptime of GCP VMs, App Engine services, website URLs, and AWS Load Balancer. Uptime checks are also a way to track the Error Budget of services. Uptime checks essentially test the availability of an external facing service within a specific timeout interval and ensure that the Error Budget of the service is not burnt unnecessarily. It is possible to initiate these tests from one or more GCP geographic regions, and a minimum of three active locations must be selected as geographic regions. Alternatively, selecting the **Global** option will initiate tests from all available locations.

The frequency at which uptime checks are performed can be configured and defaults to 1 minute. Uptime checks support multiple protocol options, such as HTTP, HTTPS, and TCP, and can be defined for the following resource types:

- **URL**: Required to specify a hostname and path.
- **App Engine Service**: Required to specify a service and path.
- **Instance**: Required to specify a path for a single instance (GCP or EC2) or a predefined group. This group needs to be explicitly configured.
- **Elastic Load Balancer**: Required to specify a path for AWS ELB.

The configuration to create an uptime check includes options to perform response validation. These options include the following:

- **Provide response timeout**: This is the time it takes for the request to complete. Must be between 1 and 60 seconds. The default is 10 seconds.
- **Enable response content**: This option allows you to select a response content match type with specific operators that contain or do not contain specific text or matches on regex.
- **Log Check Failures**: This option will save all the logs related to uptime checks failing to Cloud Logging.

In addition, alerts and notifications can be configured in situations when the uptime check fails for the selected duration. It is mandatory that the alert policy already exists and that the notification channel has been pre-created. The following screenshot shows the summary configuration of an uptime check where the target resource type is a URL:

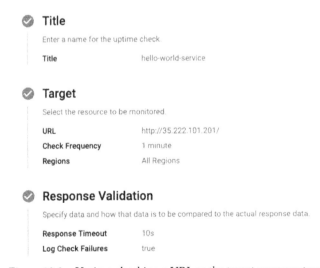

Figure 10.8 – Uptime checking a URL as the target resource type

The URL being used in the preceding screenshot is the URL of the LoadBalancer service, `hello-world-service`, that was created as part of *Chapter 8, Understanding GKE Essentials to Deploy Containerized Applications*. A configured uptime check can result in a failure. The upcoming subsection lists the potential reasons for uptime check failures.

Potential reasons for uptime check failures

The following are some potential reasons for uptime checks failing:

- **Connection errors**: The hostname/service not found or responding, or the specified port is not open or valid.

- **40x Client Errors**: Includes `403` (Forbidden Service), `404` (Incorrect Path), and `408` (port number is incorrect or service is not running).

- **Firewall rules are not configured**: If the resource being monitored by an uptime check is not publicly available, then a firewall rule needs to be configured to allow incoming traffic from uptime check servers.

> **Tip – How to identify an uptime check against service logs**
>
> Look for two specific fields in the logs: Ip field and User-agent. Ip field contains one or more addresses that are used by the uptime check server. User-agent will include some text stating `GoogleStackdriverMonitoring-UptimeChecks`.

This concludes our detailed overview of uptime checks. The next topic deep dives into alerting – a Cloud Monitoring option that's key for Incident Management. Alerting provides options for reporting on monitored metrics and providing notifications appropriately.

Alerting

Alerting is the process of processing the alerting rules, which track the SLOs and notify or perform certain actions when the rules are violated. *Chapter 3, Understanding Monitoring and Alerting to Target Reliability*, deep dived into alerting, described how alerting allows us to convert SLOs into actionable alerts, discussed key alerting attributes, and elaborated on alerting strategies. The alerting UI in Cloud Monitoring hosts information with respect to incidents currently being fired, incidents being acknowledged, active alerting policies that have been configured, details of open and closed incidents, and all the incidents tied to the events. In addition, alerting allows us to create an alert policy and configure notification channels.

Configuring an alert policy

The steps involved in configuring an alert policy are very similar to the ones for creating a chart using Metrics explorer. Essentially, an alert needs to be created against a metric. Configuring an alert includes adding a metric condition through Metrics explorer and setting a metric threshold condition.

A metric threshold condition will define the specific value. If the specific metric value falls above or below the threshold value (based on how the policy is defined), an alert will be triggered, and we will be notified through the configured notification channels. If the policy is defined through the console, then the policy trigger field is used, while if the policy is defined through the API, then the combiner field is used.

Alternatively, to define an alert policy based on a metric threshold condition, you can define an alert policy based on a metric absence condition. A metric absence condition is defined as a condition where time series data doesn't exist for a metric for a specific duration of time.

> **Important note – The alignment period is a lookback interval**
>
> The alignment period is a lookback interval from a particular point in time. For example, if the alignment period is 5 minutes, then at 1:00 P.M., the alignment period contains the samples received between 12:55 P.M. and 1:00 P.M. At 1:01 P.M., the alignment period slides 1 minute and contains the samples received between 12:56 P.M. and 1:01 P.M.

The next section describes the available notification channels that are used to send information that's specific to firing alerts.

Configuring notification channels

If an alerting policy violates the specified condition, then an incident gets created with an Open status. Information about the incident can be sent to one or more notification channels. On receipt of the notification, the operations team can acknowledge the incident through the console. This changes the status of the incident to Acknowledged. This is an indication that the event is being inspected. The incident eventually goes to Closed status if either the conditions are no longer being violated, or no data is received for the specific incident over the course of the next 7 days.

The supported notification channels are as follows:

- **Mobile Devices**: Mobile devices should be registered via the incidents section of the Cloud Console Mobile App.

- **PagerDuty Services**: Requires a service key to authenticate and authorize.

- **PagerDuty Sync**: Requires a subdomain tied to `pagerduty.com` and the respective API key to authenticate and authorize.

- **Slack**: Prompts the user to authenticate and authorize to a Slack channel through a custom URL, and then prompts the user to provide the Slack channel's name.

- **Webhooks**: Requires the endpoint URL, along with optional usage of HTTP Basic Auth.

- **Email**: Requires an email address to receive notifications when a new incident is created.

- **SMS**: Requires a phone number to receive notifications.

- **Cloud Pub/Sub**: Must specify a topic name for where the notification should be sent to. The topic should exist upfront.

> **Tip – Additional configuration to enable alert notifications to cloud Pub/Sub**
>
> When the first Pub/Sub channel is created to configure alert notifications, Cloud Monitoring will create a service account via the monitoring notification service agent, specifically in the project where the channel was created. This service account will be structured as `service-[PROJECT_NUMBER]@gcp-sa-monitoring-notification.iam.gserviceaccount.com`. The `pubsub.publisher` role should be added to the preceding service account to configure alert notifications via Cloud Pub/Sub.

This concludes this section on alerting, where we looked at configuring an alerting policy and notification channels. The next section introduces the Cloud Monitoring agent.

Monitoring agent

Cloud Monitoring provides a lot of metrics out of the box, without any additional configuration, such as CPU utilization, network traffic, and more. However, more granular metrics such as memory usage, network traffic, and so on can be collected from unmanaged VMs or from third-party applications using an optional **Monitoring agent**. The Monitoring agent is based on the `collectd` daemon (daemon refers to a program that runs in the background) to collect system statistics from various sources, including operating systems, applications, logs, and external devices.

The Monitoring agent can be installed on unmanaged GCE VMs or AWS EC2 VMs. Other Google compute services, such as App Engine, Cloud Run, and Cloud Functions, have built-in support for monitoring and do not require you to explicitly install the Monitoring agent. GKE also has built-in support for monitoring and can be enabled for new or existing clusters via *Cloud Operations for GKE*, an integrated monitoring and logging solution.

Conceptually, you must follow this process to install/configure a Monitoring agent on unmanaged VMs:

- Add the agent's package repository via a provided script that detects the Linux distribution being run on the VM and configures the repository accordingly.

- Install the Monitoring agent using the `stackdriver-agent` agent for the latest version or by using `stackdriver-agent-version-number` for a very specific version.

- Restart the agent for the installed agent to come into effect.

The step-by-step process of installing a Logging agent on a single VM/GCE VM/AWS EC2 instance can be found at `https://cloud.google.com/monitoring/agent/installation`. This completes our brief overview of the Monitoring agent. The next subsection mentions the possible access controls with respect to Cloud Monitoring.

Cloud Monitoring access controls

The following table summarizes the critical IAM roles required to access or perform actions on Cloud Monitoring:

Role Title	Role Description
Monitoring AlertPolicy Editor	Can create an alerting policy.
Monitoring Editor	To acknowledge an incident.
Monitoring Notification Channel Editor	To configure a notification channel
Monitoring Dashboard Viewer	Can view dashboard settings
Monitoring Dashboard Editor	Can create a dashboard or edit dashboard settings.
Monitoring Metric Writer	Allows you to write monitoring data to a workspace but doesn't permit you to view it via the console. Used only by service accounts.
Monitoring Viewer	Read-only access to the console and API.
Monitoring Editor	Read-write access to the console and API. Allows you to write monitoring data to a workspace.
Monitoring Admin	Full access to all monitoring features. Allows you to create a workspace.

Groups – A collection of resources that is defined as a Monitoring Group

> **Note**
>
> Cloud Monitoring allows you to create a monitoring group. This is a convenient way to view the list of GCP resources, events, incidents, and visualizations as key metrics from a centralized place. A monitoring group is created by defining one or more criteria either against the name, resource type, tag, security group, cloud project, or region. If multiple criteria are specified, then an OR/AND operator can be specified.

This concludes our deep dive into Cloud Monitoring and its respective constructs, such as Workspace, dashboards, Metrics explorer, uptime checks, alerting policies, and access controls. The next section elaborates on another GCP construct that's part of Cloud Operations and focuses on logging; that is, Cloud Logging.

Cloud Logging

A log is defined as a record of a status or event. Logging essentially describes what happened and provides data so that we can investigate an issue. It is critical to be able to read and parse logs across a distributed infrastructure involving multiple services and products. **Cloud Logging** is a GCP service that allows you to store, search, analyze, monitor, and alert others about logging data and events from Google Cloud and AWS, third-party applications, or custom application code. The information in the log entry is structured as a payload. This payload consists of information related to a timestamp, a resource that the log entry applies to, and a log name. The maximum size of a log entry is 256 KB. Each log entry indicates the source of the resource, labels, namespaces, and status codes. Cloud Logging is also the source of input for other Cloud Operations services, such as Cloud Debug and Cloud Error Reporting.

The following are the key features of Cloud Logging:

- **Audit Logs**: Logs are captured and categorized as Admin Activity, Data Access, System Event, and Access Transparency Logs, with each category having a default retention period.

- **Logs Ingestion**: Logs can be ingested from many sources, including GCP services and on-premises or external cloud providers, by using the Cloud Logging API or through logging agents.

- **Logs Explorer**: Logs can be searched for and analyzed through a guided filter configuration or flexible query language, resulting in effective visualization. Results can also be saved in JSON format.

- **Logs-based Metrics**: Metrics can be created from log data and can be added to charts/dashboards using the Metrics explorer.

- **Logs Alerting**: Alerts can be created based on the occurrence of log events and based on the created logs-based metrics.

- **Logs Retention**: Logs can be retained for a custom retention period based on user-defined criteria.

- **Logs Export**: Logs can be exported to Cloud Storage for archival, BigQuery for advanced analytics, Pub/Sub for event-based processing using GCP services, user-defined cloud logging sinks, or to initiate external third-party integrations so that you can export using services such as Splunk.

Cloud Logging features will be discussed in detailed in the upcoming subsections, starting with Audit Logs.

Audit Logs

Cloud Audit Logs is a fundamental source for finding out about certain parts of a project (*who did what, where, and when?*). Cloud Audit Logs maintains logs for each project (including folder- and organization-level information). Cloud Audit Logs can be categorized into various categories. Let's take a look.

Admin activity logs

Admin activity logs are specific to any administrative actions that modify the configuration or metadata of resources. Examples for admin activity logs include, but not are limited to, the following:

- Setting or changing permissions of a cloud storage bucket

- Assigning /unassigning IAM roles

- Changing any properties of a resource, such as tags/labels

- Creating/deleting resources for GCE, GKE, or Cloud Storage

The following screenshot shows the admin activity logs for when a GCE VM or bucket was created. The easiest way to access these activity logs is from the **Activity** tab on the GCP console home page. This pulls a live feed of the admin activity logs but does not include data access logs by default:

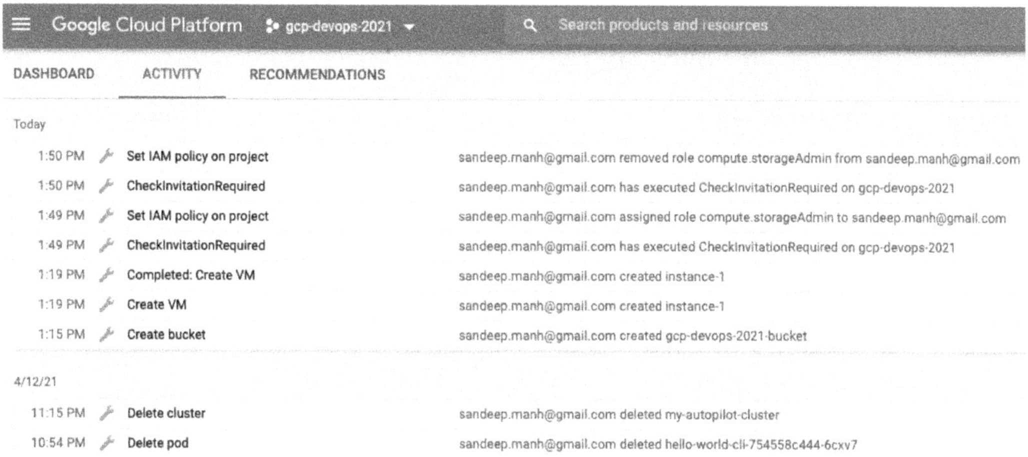

Figure 10.9 – Admin activity logs

The next subsection provides an overview of an audit log category specific to *data access*.

Data access logs

Data access logs are useful for reading the configuration or metadata of resources. This also includes user-level API calls, which read or write resource data. Data access audit logs need to be enabled explicitly (except for Big Query), and this can be controlled by specifying the services whose audit logs should be captured. In addition, data access logs tied to actions performed by a specific set of users or groups can be exempted, thus providing granular control. Data access logs can be further classified into three subtypes:

- **Admin read**: Read attempts on service metadata or configuration data. An example of this is listing the available buckets or listing the nodes within a cluster.

- **Data read**: Read attempts of data within a service. An example of this includes listing the objects within a bucket.

- **Data write**: Write attempts of data to a service. An example of this includes creating an object within a bucket.

The following is a screenshot of granular data access being configured for an individual GCP service from IAM – the **Audit Logs** UI:

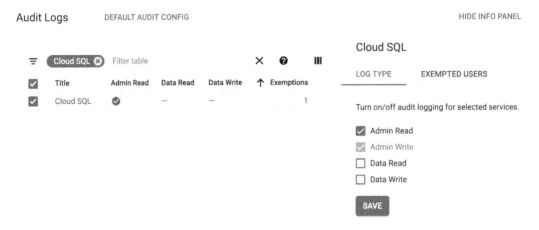

Figure 10.10 – Configuring IAM Audit Logs for a service

The preceding screenshot also shows the option to configure **Exempted Users**. This option allows you to exempt audit logs from being generated for certain users, as configured. Data access logs can be viewed either through the **Activity** tab of the GCP home page, where **Activity Type** is **Data Access**, or through the **Logs Explorer** UI (discussed later). The next subsection provides an overview of an audit log category specific to system events.

System event logs

System event logs are used when changes have been made to resources by Google systems or services. They are not specific to user actions on the resources. Examples of system event logs include, but are not limited to, the following:

- Automatically restarting or resetting Compute Engine

- System maintenance operations, such as migration events, which are performed by Compute Engine to migrate applications to a different host

The next subsection provides an overview on an audit log category specific to access transparency.

Access transparency logs

Access transparency logs are used by Google personnel when they're accessing a user's/customer's content. This situation typically arises when Google's support team is working on a customer issue (such as a specific service not working as expected or an outage) and, as a result, needs to access the customer's project. This category of logs is critical if you wish to follow legal and regulatory obligations. In addition, you can trace events to look back on the actions that have been performed by Google support personnel. Access transparency logs can be enabled by contacting Google support and are available for customer support levels, excluding individual accounts. An example of access transparency logs could be the logs that are accessed by the support personnel while trying to resolve a support issue for a VM instance.

Policy denied logs

Policy denied logs are logs that are captured when access is denied by a Google Cloud service to either a user or service account. Policy denied logs can be excluded from ingestion into Cloud Logging through Logs Exclusions.

This completes this section on audit logs, where we provided an overview of the various subcategories. Before proceeding to the next section, take a look at the following table, which lists the IAM roles specific to accessing logs:

Role Name	Role Description
Logs Viewer and Project Viewer	View logs except Data Access/Access Transparency logs, as they might contain sensitive data.
Project Editor	Write, view, and delete logs but cannot view Data Access/Access Transparency logs.
Private Logs Viewer	View all types of Audit Logs, including Data Access and Access Transparency Logs.
Logging Admin and Project Owner	Full access to all logging actions.

The next section will explain how logs are ingested into Cloud Logging from multiple sources.

Logs ingestion, routing, and exporting

Cloud Logging supports logs ingestion from multiple sources, such as audit logs, service logs, application logs, syslogs, and platform logs. These logs are sent to the **Cloud Logging API**. The Cloud Logging API forwards the incoming log entries to a component called **Logs Router**. Logs Router is fundamentally responsible for routing logs to their respective destinations.

These destinations can be grouped into four possible categories. Logs Router will check the incoming logs against existing rules to determine whether to ingest (store), export, or exclude them, and will route the logs to one of the four destination categories.

These destination categories are as follows:

- **_Required log bucket**: This is the primary destination for admin activity, system event, and access transparency logs. There are no charges associated with these logs and this bucket cannot be modified or deleted.

- **_Default log bucket**: This is the primary destination for data access, policy denied, and user-specific logs. There are charges associated with these logs. The bucket cannot be deleted but the _Default log sink can be disabled.

- **User-managed log sinks**: A user-managed **log sink** is an object that holds the filter criteria and a destination. The destination could either be Cloud Storage, Cloud Logging, BigQuery, Pub/Sub, or Splunk. The user-managed log sink is the destination for incoming logs from the Cloud Logging API that satisfy the filter criteria defined against the sink. These are also known as **Inclusion Filters**. This applies to logs that fall under the _Required log bucket and the _Default log bucket. The process of writing logs to user-managed log sinks can also be characterized as **Log Exports** (if the intent is to export for external processing) or **Log Retention** (if the intent is to export to retain logs for a longer period from a compliance perspective).

- **Exclusions**: Exclusions are governed by log exclusion filters. They only apply to entries that qualify for the _Default log bucket. In other words, logs that qualify for the _Required log bucket can never be excluded. If any of the log exclusion filters match with entries that qualify for the _Default log bucket, then the entries will be excluded and never be saved.

> **Tip – What are log buckets?**
>
> Log buckets are a form of object storage in Google Cloud projects that are used by Cloud Logging to store and organize logs data. All logs generated in the project are stored in these logs' buckets. Cloud Logging automatically creates two buckets in each project: _Required and _Default._ Required represents an audit bucket, which has a 400-day retention period, while _Default represents the *everything else* bucket, which has a 30-day retention period. In addition, a user can create custom logging buckets, also known as user-managed log sinks.
>
> The log bucket per project can be viewed via the **Logs Storage UI** in Cloud Logging. Additional actions such as creating a user-defined log bucket and a usage alert can also be initiated from the Logs Storage UI.

The following diagram illustrates how logs are ingested from multiple sources via the Cloud Logging API and, subsequently, routed by Logs Router to possible destinations:

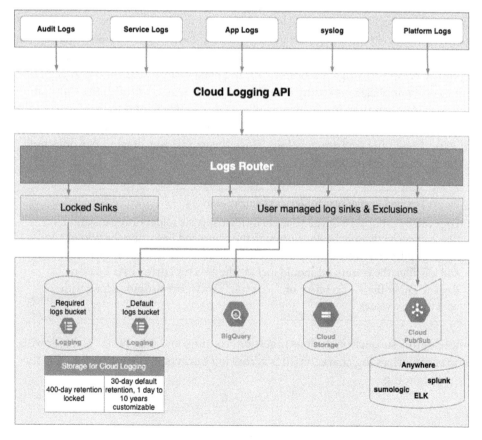

Figure 10.11 – Illustrating logs ingesting and logs routing

There are three steps we must follow to export logs:

1. Create a sink.

2. Create a filter that represents the criteria to identify logs to export.

3. Create a destination – Cloud Storage Bucket, BigQuery, or Pub/Sub topics.

> **IAM roles to Create/Modify/View a Sink**
>
> The Owner or Logging/Logs Configuration Writer role is required to create or modify a sink. The Viewer or Logging/Logs Viewer role is required to view existing sinks. The Project Editor role does not have access to create/edit sinks.

To summarize, logs can originate from multiple sources, such as on-premises, Google Cloud, or a third-party cloud service. These logs are injected into Cloud Logging through the Cloud Logging API, which are then sent to Logs Router. Logs Router, which is based on configured filters, will route the logs to logged sinks (the _Required or _Default log bucket). Additionally, a copy of the logs can be sent to user-managed sinks based on the configured filter criteria, where the destination can either be Cloud Storage, BigQuery, or Pub/Sub. Log export can be used for multiple purposes, such as long-term retention for compliance reasons (using Cloud Storage), Big Data analysis (using BigQuery), or to stream to other applications (using Pub/Sub). If these logs are sent to the Pub/Sub messaging service, then they can be exported outside Google Cloud to third-party tools such as Splunk, Elastic Stack, or SumoLogic. It is important to note that configured log sinks for export will only capture new logs, since the export was created but does not capture the previous logs or backfill.

> **How to export logs across folders/organizations**
>
> Logs can be exported from all projects inside a specific folder or organization. This can currently only be done through the command line using the gcloud logging sink's `create` command. Apart from the sink's name, destination, and log filter, the command should include the `--include-children` flag and either the `--folder` or `--organization` attribute, along with its respective values.

This completes this subsection on logs ingestion, routing, and exporting. The following subsection summarizes log characteristics across log buckets in the form of a table for ease of understanding.

Summarizing log characteristics across log buckets

Each log type is destined to a specific Cloud Logging bucket. In addition, every log type has specific characteristics in terms of the minimum IAM roles required to access the logs, the default retention period, and the ability to configure a custom retention period. The following table details the respective information:

Log Type	IAM roles (Min)	Default Retention Period	Custom Retention	Logs Bucket
Admin Activity	Logging/Logs Viewer	400 days	Not configurable	_Required
System Event	Logging/Logs Viewer	400 days	Not configurable	_Required
Access Transparency	Logging/Private Logs Viewer	400 days	Not configurable	_Required
Data Access	Logging/Private Logs Viewer	30 days	Configurable	_Default
Policy Denied	Logging/Logs Viewer	30 days	Configurable	_Default
All other logs	Logging/Logs Viewer	30 days	Configurable	_Default

> **All other logs**
>
> This refers to either user logs generated by applications through a Logging agent or platform logs generated by GCP services or VPC Flow Logs or Firewall Logs.

In addition to the preceding table, it is important to note the following:

- System event logs are system initiated, whereas admin activity, data access, and access transparency logs are user initiated.
- Admin activity and system event logs record the changes in the configuration of resources, whereas data access logs record the changes that were made inside the record.
- Admin activity, system event, and access transparency logs are always enabled.

This completes our overview on logs ingestion. The next topic focuses on the **Logs Explorer** UI, through which users can explore ingested logs.

Logs Explorer UI

Logs Explorer UI is the centralized way to view logs that have been ingested into Cloud Logging via the Cloud Logging API, and ultimately routed via Cloud Router to either Cloud Logging buckets or user-managed sinks. The UI allows us to filter logs by writing advanced search queries, visualize the time series data by configuring time windows, and perform critical actions to create log-based metrics or create users. The UI consists of multiple options and sections, as shown in the following screenshot:

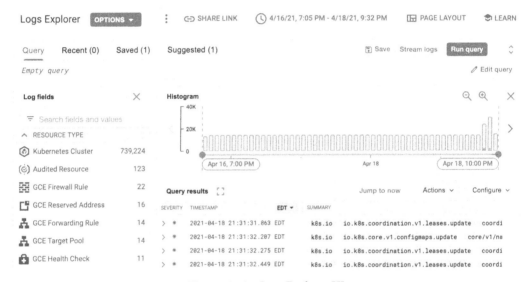

Figure 10.12 – Logs Explorer UI

To filter Cloud Audit Logs through the Logs Explorer UI, select the following options for the **Log Name** field:

- **Admin Activity**: `cloudaudit.googleapis.com%2Factivity`

- **Data Access**: `cloudaudit.googleapis.com%2Fdata_access`

- **System Event**: `cloudaudit.googleapis.com%2Fsystem_event`

Let's take a look at some key important information with respect to navigating the options in the Logs Explorer UI.

Query builder

This section constructs queries to filter logs. Queries can be expressed in query builder language by choosing an appropriate combination of field and value, as shown in the following screenshot. The user can provide input in two ways:

- By choosing options from the available drop-down menus with respect to **Resource**, **Log Name**, and **Severity**. This is the basic query interface.

- By choosing fields from the **Log Fields** section, starting by either selecting the **Resource** type or the **Severity** type. This is the advanced query interface:

Figure 10.13 – Query builder section under Logs Explorer

Query results

This section displays the results that match the filter criteria defined within query builder. If there is a match, the results are displayed in one or more rows. Each row represents a log entry, as shown here:

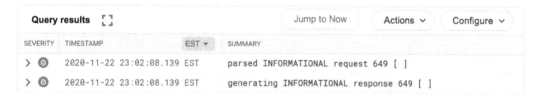

Figure 10.14 – Query results section

Log entries

Each query result that's returned is a log entry that's displayed with a timestamp and summary text information. When expanded, the log entry displays further details in a JSON payload format. The JSON payload has multiple fields and can be elaborated on using the **Expand nested fields** option. Additionally, the user can copy the payload to a clipboard or share the specific payload by copying the shareable link, as shown here:

Figure 10.15 – Viewing the JSON payload for a log entry

Payload-specific actions

There are multiple options that perform actions on a specific payload on a specific field, as shown in the following screenshot. These are as follows:

- **Show matching entries**: Adds the selected key-value pair from the JSON payload to the existing filter criteria and shows matching entries within the configured time window.

- **Hide matching entries**: Adds the selected key-value pair from the JSON payload to the existing filter criteria in a negation form and removes the matching entries from user display, within the configured time window.

- **Add field to summary line**: Adds the selected key to the summary section:

Figure 10.16 – Possible payload-specific actions for a specific field

Page layout

This option allows users to configure the page layout and optionally include **Log Fields** and/or a **Histogram**. **Query builder** and **Query results** are mandatory sections and cannot be excluded:

Figure 10.17 – Options under the PAGE LAYOUT section

Actions (to perform on a query filter)

Actions allows user to operate on the potential results from the query filter definition. This includes **Create Metrics**, **Download Logs**, and **Create Sinks**:

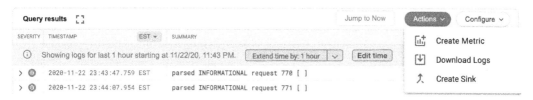

Figure 10.18 – Possible actions you can perform on a query filter

This completes this section on Logs Explorer and all the possible UI options for filtering and analyzing logs. The next section provides an overview of *logs-based metrics*.

Logs-based metrics

Logs-based metrics are Cloud Monitoring metrics that are created based on the content of the log entries. They can be extracted from both included and excluded logs. As matching log entries are found, the information that's tied to the metrics is built over time. This forms the required time series data that is critical to metrics. Logs-based metrics are used in creating Cloud Monitoring charts and can also be added to Cloud Monitoring dashboards.

> **Important note**
> To use logs-based metrics, a Google Cloud project is required with billing enabled. In addition, logs-based metrics are recorded for matching log entries, once the metric has been created. Metrics are not backfilled for log entries that are already in Cloud Logging.

Logs-based metrics can be classified as either of the following:

- System (logs-based) metrics
- User-defined (logs-based) metrics

Both of these logs-based metrics will be covered in the upcoming subsections.

System (logs-based) metrics

System (logs-based) metrics are out-of-the-box, predefined metrics from Google and are very specific to the current project. These metrics record the number of events that occur within a specific period. A list of available system (logs-based) metrics can be found under the **Logs-based Metrics** UI in the **Logging** section of Cloud Operations. Examples include the following:

- **byte_count**: Represents the total number of received bytes in log entries
- **excluded_byte_count**: Represents the total number of excluded bytes from log entries

The user can create an alert from a predefined metric or view the details of the metric, along with its current values, in Metrics explorer:

Figure 10.19 – System (logs-based) metrics and their qualifying actions

The next section provides an overview of user-defined metrics.

User-defined (logs-based) metrics

User-defined (logs-based) metrics, as the name suggests, are defined by the user and are specific to the project where the user configures these metrics. These metrics can either of the **Counter** or **Distribution** type:

- **Counter**: Counts the number of log entries that match on a query

- **Distribution**: Accumulates numeric data from log entries that match on a query

Users can create a user-defined metric either from **Logs-based Metrics UI** via the **Create Metric** action or the **Logs Explorer UI** via the actions menu above the query results. Once the user initiates these actions, they get to choose the type of metric in the Metric Editor panel; that is, **Counter** or **Distribution**.

In addition, the user will have to configure fields such as the metric's name, description, and any optional labels and units. For a **Counter,** the units should be left blank. For a **Distribution**, the units should be s, ms, and so on:

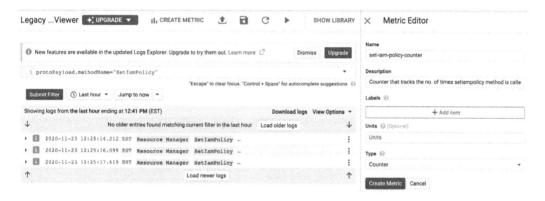

Figure 10.20 – Creating a logs-based metric

More details on creating a distribution metric can be found at `https://cloud.google.com/logging/docs/logs-based-metrics/distribution-metrics`.

Access control for logs-based metrics

The following table displays the critical IAM roles required, along with their minimal permissions (in accordance with the principle of least privilege), to access or perform actions related to logs-based metrics:

Role Name	Role Description
Logs Configuration Writer	Create, update, and delete logs-based metrics, buckets, views, and export sinks.
Logs Viewer	View existing metrics.
Monitoring Viewer	Read the time series that the logs-based metrics contains.
Project Viewer	Similar to Logs Configuration Writer but cannot create or export sinks, nor view Data Access/Access Transparency logs.

The following section will conclude this section on Cloud Logging by exploring the available network-based log types on Google Cloud.

Network-based log types

There are two network-based log types that primarily capture logs related to network interactions. These are as follows:

- VPC Flow Logs
- Firewall logs

Let's look at them in detail.

VPC Flow Logs

VPC Flow Logs capture real-time network activity (incoming/outgoing) against VPC resources on an enabled subnet. Flow logs capture activity specific to the TCP/UDP protocols and are enabled at a VPC subnet level. Flow logs generate a large amount of chargeable log files, but they don't capture 100% of traffic; instead, traffic is sampled at 1 out of 10 packets and cannot be adjusted. Flow logs are used for Network Monitoring – to understand traffic growth from a forecasting capacity and for forensics – to evaluate network traffic (in/out) in terms of traffic source. Flow logs can be exported for analysis using BigQuery. In the case of a Shared VPC – where multiple service projects connect to a common VPC – flow logs flow into the host project, not the service projects.

Firewall logs

Firewall logs capture the effects of a specific firewall rule in terms of the traffic that's allowed or denied by that firewall rule. Similar to VPC Flow Logs, firewall logs capture TCP/UDP traffic only and are used for auditing, verifying, and analyzing the effect of the configured rules. Firewall logs can be configured for an individual firewall rule. Firewall rules are applied for the entire VPC and cannot be applied at a specific subnet level like flow logs. Firewall logs attempt to capture every firewall connection attempt on a best effort basis. Firewall logs can also be exported to BigQuery for further analysis.

Every VPC has a set of hidden implied pre-configured rules, with the lowest priority being 65535. Firewall rules can have a priority between 0 and 65535 (0 implies highest, while 65535 implies lowest). These are as follows:

- `deny all ingress`: By default, this denies all incoming traffic to the VPC.
- `allow all egress`: By default, this allows all outgoing traffic from the VPC.

However, firewall logs cannot be enabled for the hidden rules. So, to capture the incoming traffic that is being denied or the outgoing traffic that is being allowed, it is recommended to explicitly configure a firewall rule for the denied/allowed traffic with an appropriate priority and enable firewall logs on that rule.

This completes this subsection on network-based log types, where we introduced VPC Flow Logs and firewall logs.

Logging agent

The **Logging agent** is optional and is used to capture additional VM logs, such as **Operating System (OS)** logs (such as Linux syslogs or Windows Event Viewer logs) and logs from third-party applications. The Logging agent is based on `fluentd` – an open source log or data collector. The Logging agent can be installed on unmanaged GCE VMs or AWS EC2 VMs. Other Google Compute services such as App Engine, Cloud Run, and Cloud Functions have built-in support for logging and do not require you to explicitly install the Logging agent. GKE also has built-in support for logging and can be enabled for new or existing clusters by *Cloud Operations for GKE*, an integrated monitoring and logging solution.

To configure the Logging agent, you must configure an additional configuration file, but a single configuration file acts as a catch all for capturing multiple types of logs, including OS logs and third-party application logs such as Apache, MySQL, Nginx, RabbitMQ, and so on. However, there are scenarios where the configuration file of the agent needs to be modified so that we can modify the logs. These are as follows:

- When reformatting log fields, either the order or combine multiple fields into one

- When removing any **Personally Identifiable Information** (**PII**) or sensitive data

- When modifying records with `fluentd` plugins such as `filter_record_transformer`, a plugin for adding/modifying/deleting fields from logs before they're sent to Cloud Logging

Conceptually, the following is the process of installing/configuring an agent on a GCE VM:

1. Add the agent's package repository via a provided script that detects the Linux distribution being run on the VM and configures the repository accordingly.

2. Install the Logging agent and install the `google-fluentd-catch-all-config` agent for unstructured logging and the `google-fluentd-catch-all-config-structured` agent for structured logging.

3. Restart the agent for the installed agents to come into effect.

The step-by-step process of installing a Logging agent on a single VM/GCE VM/AWS EC2 instance can be found at `https://cloud.google.com/logging/docs/agent/installation`.

This completes our high-level overview of logging agents. Subsequently, this also completes the section on Cloud Logging, where we looked at features such as audit log types, logs ingestion, the Logs Explorer UI, logs-based metrics, and access controls. The next section deep dives into *Cloud Debugger*, a GCP construct from Cloud Operations that can potentially inspect a production application by taking a snapshot of it, without stopping or slowing down.

Cloud Debugger

Cloud Debugger allows us to inspect the state of a running application in real time. Cloud Debugger doesn't require the application to be stopped during this process and doesn't slow it down, either. Users can capture the call stack and variables at any location in the source code. This essentially allows the user to analyze the application state, especially in complex situations, without adding any additional log statements.

In addition, Cloud Debugger can be used for production environments and is not limited to development or test environments. When Cloud Debugger captures the application state, it adds request latency that is less than 10 ms, which, practically, is not noticeable by users.

Cloud Debugger is supported on applications running in GCP such as App Engine, Compute Engineer, GKE, Cloud Run, and so on, as well as those written in a number of languages, including Java, Python, Go, Node.js, Ruby, PHP, and .NET. Cloud Debugger needs access to the application code and supports reading the code from App Engine, Google Cloud source repositories, or third-party repositories such as GitHub, Bitbucket, and so on.

Setting up Cloud Debugger

Enabling/setting up Cloud Debugger involves the following fundamental steps:

1. Enable the Cloud Debugger API as a one-time setup per project.

2. Provide appropriate access so that the GCP service where Cloud Debugger will run has permission to upload telemetry data or call Cloud Debugger.

3. App Engine and Cloud Run must already be configured for Cloud Debugger.

4. A service account with the Cloud Debugger Agent role is required for applications running in Compute Engine, GKW, or external systems.

5. If the application is running inside a Compute Engine VM or cluster nodes with a default service account, then the following access scopes should be added to the VMs or cluster nodes: `https://www.googleapis.com/auth/cloud-platform` and `https://www.googleapis.com/auth/cloud_debugger`.

6. Select the source code location. If there is no access to the source code, a debug snapshot can be taken that captures the call stack and local variables.

7. If there is access to the source code, then App Engine standard will select the source code automatically. App Engine flex, GCE, GKE, and Cloud Run can automatically select the source code based on the configuration file in the application root folder; that is, `source-context.json`.

8. Alternatively, select a source code location from the possible options, including local files, Cloud Source Repositories, GitHub, Bitbucket, and GitLab.

9. To enable Cloud Debugger from application code, you must follow a set of instructions that are specific to the language that the application has been written in. The following is an example snippet:

```
try:
    import googleclouddebugger
    googleclouddebugger.enable()
except ImportError:
    pass
```

Now that we've set up Cloud Debugger, let's learn how to use it.

Using Cloud Debugger

Using Cloud Debugger involves learning about the functionality of debug snapshots, debug logpoints, and accessing the logs panel.

Debug snapshots

Snapshots capture local variables and the call stack at a specific location in the application's source code. The fundamental step prior to taking a snapshot is to set up a breakpoint. It takes about 40 seconds for a breakpoint to come into effect. Cloud Debugger breakpoints do not stop code execution. A non-intrusive snapshot is taken when the flow of execution passes the debug point. Additional conditions can be added so that a snapshot is only taken if a data condition passes. The captured snapshot will contain details of the local variables and the state of the call stack.

In the following screenshot, the breakpoint was set to line 39 against a specific file. The breakpoint has a qualifying condition, and a snapshot is taken if its condition is met. The details of the variables are displayed in the **Variables** section:

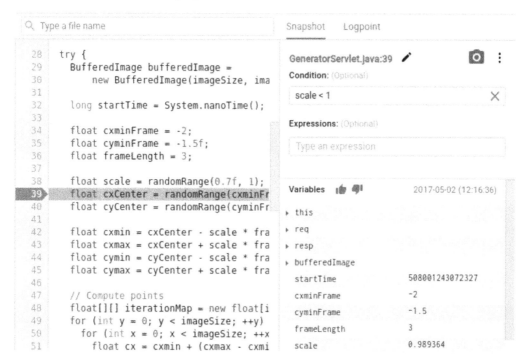

Figure 10.21 – Taking a debug snapshot in Cloud Debugger

Optionally, expressions can also be included while configuring a snapshot. Expressions can be used as special variables to evaluate values when a snapshot is taken. These are especially useful in scenarios where the values being captured by the expressions are not usually captured by local variables.

In the following screenshot, we can see that multiple expressions are defined while configuring a snapshot and are captured while taking a snapshot:

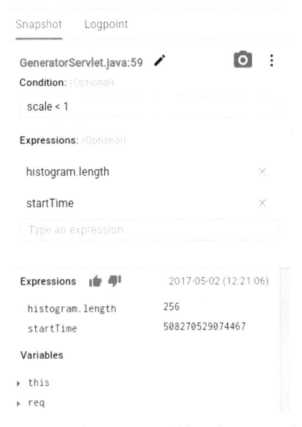

Figure 10.22 – Defining expressions while configuring a snapshot

The following are some key pointers related to snapshots:

- A snapshot is taken only once. To capture another snapshot of the application data for the same location in the code, the user needs to manually retake the snapshot through the camera icon in the snapshot panel.

- A snapshot location can be manually removed by clicking the **x** icon on the breakpoint.

- Cloud Debugger generates a new URL for every snapshot that's been taken. It is valid for 30 days from the time it was taken. This URL can be shared with other members of the project.

The next subsection provides an overview of debug logpoints and how they can be injected into a running application.

Debug logpoints

It's a common practice to add log messages when you're trying to solve complex problems. In such scenarios, developers often provide code changes to production that essentially include additional log statements that help with analysis. If the problem is complex, this process needs to be repeated multiple times, which means the production code needs to go through multiple changes to include log statements. Cloud Debugger steps away from the traditional approach to debugging an application and instead provides a dynamic way to add log messages using **debug logpoints**.

Debug logpoints can inject logs into a running application without stopping, editing, or restarting. A logpoint is added at a location of choice, as per the developer's wishes. When that particular portion of code is executed, Cloud Debugger logs a message, and the log message is sent to the appropriate service that is hosting the application. So, if the application is hosted in App Engine, then the log message can be found in the logs tied to App Engine.

In the following screenshot, a logpoint has been added with a condition, with the message log level set to **Info**. The concept of specifying a condition along with a logpoint is called a *logpoint condition*. This is an expression in the application language that must evaluate to true for the logpoint to be logged. Logpoint conditions are evaluated each time that specific line is executed, if the logpoint is valid:

Figure 10.23 – Adding a debug logpoint via Cloud Debugger

The following are some key pointers related to logpoints:

- A logpoint can be created even if direct access to the source code is not available. A logpoint can be created by specifying the name of the file, the line number to create the logpoint, the log level, an optional condition, and an appropriate message, as shown here:

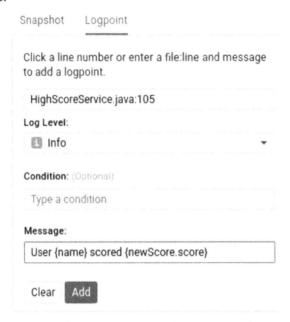

Figure 10.24 – Configuring a logpoint without access to the source code

- Logpoints becomes inactive after 24 hours and post that, messages with respect to those logpoints will not be evaluated or logged.

- Logpoints are automatically deleted after 30 days from the time of creation. Optionally, users can manually delete logpoints at will.

The next subsection illustrates the usage and options available in the *Logs panel*.

Logs panel

Cloud Debugger includes an in-page Logs panel that displays the running logs of the current application being inspected. This allows the developer to view logs next to the respective code. Users can use the logs panel to perform search variations, including text-based search, and can filter by either log level, request, or file. The results are highlighted in the context or are shown in the logs viewer:

Figure 10.25 – Logs panel for viewing logs while debugging in Cloud Debugger

The upcoming subsection provides an overview of the access control that's required for Cloud Debugger.

Access control for Cloud Debugger

The following table displays the critical IAM roles required, along with their minimal permissions (in accordance with the principle of least privilege), to access or perform actions related to Cloud Debugger:

Role Name	Role Description
Cloud Debugger Agent	Register the debug target, read active breakpoints, and report breakpoint results. This role is normally assigned to the service account running with the Debugger agent.
Cloud Debugger User	Create, view, list, and delete breakpoints (snapshots and logpoints), as well as list debug targets.

> **Tip – How to hide sensitive data while using debugging**
>
> Cloud Debugger has a feature in Pre-GA where sensitive data can be hidden through a configuration file. This configuration file consists of a list of rules that are either expressed as `blacklist` or `blacklist_exception` (to specify an inverse pattern). If the criteria match, then data is hidden and is reported by the debugger as `blocked by admin`. This feature is currently only supported for applications written in Java.

This completes this section on Cloud Debugger, where we learned how to set up Cloud Debugger, utilize debug logpoints to add log messages, and create snapshots to capture the call stack and its local values. We looked at the options that are available in the Logs panel and looked at the required access controls we can use to perform actions related to Cloud Debugger. In the next section, we will look at *Cloud Trace*, another GCP service that is part of Cloud Operations. Cloud Trace represents a distributed tracing system that collects latency data from applications to identify bottlenecks.

Cloud Trace

A **trace** is a collection of spans. A **span** is an object that wraps latency-specific metrics and other contextual information around a unit of work in an application. **Cloud Trace** is a distributed tracing system that captures latency data from an application, tracks the request's propagation, retrieves real-time performance insights, and displays the results in Google Cloud Console. This latency information can be either for a single request or can be aggregated for the entire application. This information helps us identify performance bottlenecks.

Additionally, Cloud Trace can automatically analyze application traces that might reflect recent changes to the application's performance, identify degradations from latency reports, capture traces from containers, and create alerts as needed.

Cloud Trace's language-specific SDKs are available for Java, Node.js, Ruby, and Go. These SDKs can analyze projects running on VMs. It is not necessary for these VMs to only be running on Google Cloud. Apart from the SDK, the Trace API can be used to submit and retrieve trace data from any source. A Zipkin collector is available, which allows Zipkin tracers to submit data to Cloud Trace. Additionally, Cloud Trace can generate trace information using OpenCensus or OpenTelemetry instrumentation. Cloud Trace consists of three main sections: Trace Overview, Trace List, and Analysis Reports. Let's look at them in detail.

Trace Overview

The **Trace Overview** page provides a summary of latency data that's spread across various informational panes:

- **Insights**: Displays a list of performance insights, if applicable

- **Recent Traces**: Highlights the most recent traces for a project

- **Frequent URIs**: Displays a list of URIs along with their average latency for the most frequent requests to the application in the last 7 days

- **Frequent RPCs**: Displays a list of RPCs along with their average latency for the most frequent RPC calls made in the last 7 days

- **Chargeable trace spans**: Summarizes the number of trace spans that have been created and received by Cloud Trace for the current and previous months:

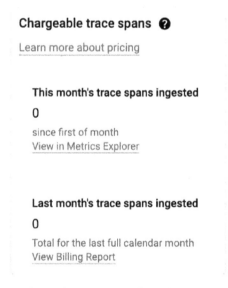

Figure 10.26 –Chargeable trace spans from the Trace Overview page

The next subsection provides an overview of the **Trace List** window, which can be used to examine traces in detail.

Trace List

The **Trace List** window allows users to find, filter, and examine individual traces in detail. These traces are displayed in a heatmap, and a specific section of the heatmap can be selected if you wish to view these traces within that specific slice of the window:

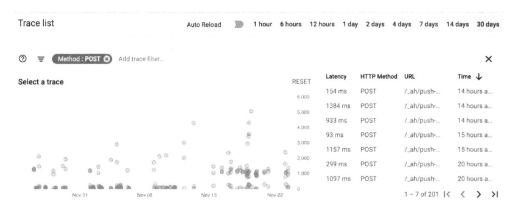

Figure 10.27 – List of all the traces filtered by the POST method in the last 30 days

Clicking on the individual trace (represented by a circle) provides details about the trace. It is represented by a waterfall graph:

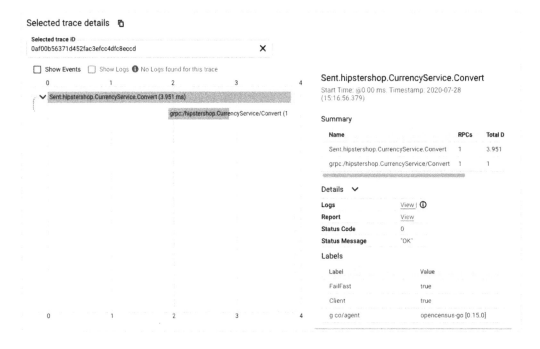

Figure 10.28 – Waterfall graph of an individual trace

The next subsection provides an overview of trace analysis reports with respect to request latency.

Analysis Reports

Analysis Reports shows an overall view of the latency for all the requests or a subset of requests with respect to the application. These reports are categorized either as daily reports or custom analysis reports.

Daily reports

Cloud Trace creates a daily report automatically for the top three endpoints. Cloud Trace compares the previous days' performance with the performance from the same day of the previous week. The content of the report cannot be controlled by the user.

Custom analysis reports

The user can create a custom analysis report, where the content of the report can be controlled from the aspect of which traces can be included. The report can include latency data either in histogram format or table format, with links to sample traces. The report can optionally include a bottleneck pane that lists **Remote Procedure Calls** (**RPCs**), which are significant contributors to latency.

> **Condition to auto-generate or manually create a trace report**
>
> For the daily report to auto-generate or for a user to create a custom report within a specific time range, it is mandatory that at least 100 traces are available in that time period. Otherwise, a trace report will not be generated.

This completes this section on Cloud Trace, a GCP construct for representing a distributed tracing system, collecting latency data from applications, and identifying performance bottlenecks. The next section focuses on Cloud Profiler, a service that is part of Cloud Operations. Cloud Profiler is a low-impact production profiling system that presents call hierarchy and resource consumption through an interactive flame graph.

Cloud Profiler

Cloud Profiler provides low-impact continuous profiling to help users understand the performance of a production system. It provides insights into information such as CPU usage, memory consumption, and so on. Cloud Profiler allows developers to analyze applications running either in Google Cloud, other cloud providers, or on-premises.

Cloud Profiler uses statistical techniques and extremely low-impact instrumentation to provide a complete picture of an application's performance, without slowing it down. Cloud Profiler runs across all production application instances, presents a call hierarchy, and explains the resource consumption of the relevant function in an interactive flame graph. This information is critical for developers to understand which paths consume the most resources and illustrates the different ways in which the code is actually called. The supported programming languages include Java, Go, Node.js, and Python.

Cloud Profiler supports the following types of profiles:

- **CPU time**: The time the CPU spent executing a block of code. This doesn't include the time the CPU was waiting or processing instructions for something else.

- **Heap**: Heap or heap usage is the amount of memory that's allocated to the program's heap when the profile is collected.

- **Allocated heap**: Allocated heap or heap allocation is the total amount of memory that was allocated in the program's heap, including memory that has been freed and is no longer in use.

- **Contention**: Contention provides information about the threads that are stuck and the ones waiting for other threads. Understanding contention behavior is critical to designing code and provides information for performance tuning.

- **Threads**: Information related to threads gives insights into the threads that are created but never actually used. This forms the basis for identifying leaked threads, where the number of threads keeps increasing.

- **Wall time**: Wall time is the time it takes to run a block of code, including its wait time. The wall time for a block of code can never be less than the CPU time.

The following screenshot summarizes the supported profile types by language:

Profile type	Go	Java	Node.js	Python
CPU time	Y	Y		Y
Heap	Y	Y	Y	
Allocated heap	Y			
Contention	Y			
Threads	Y			
Wall time		Y	Y	Y

Figure 10.29 – Supported profile types by language

The following screenshot shows the Profiler interface, which depicts a sample interactive flame graph for the **CPU time** profile type. The profile data is retained for 30 days and the profile information can be downloaded for long-term storage:

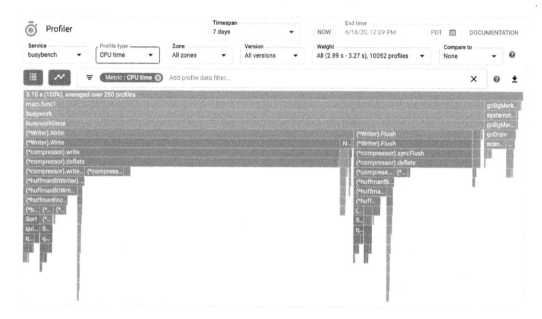

Figure 10.30 – Interactive flame graph with the profile type set to CPU time

The upcoming subsection explains the access controls that are required to perform actions with respect to Cloud Profiler.

Access control for Cloud Profiler

The following table displays the critical IAM roles required, along with their minimal permissions (in accordance with the principle of least privilege), to access or perform actions related to Cloud Profiler:

Role Name	Role Description
Cloud Profiler Agent	Register and provide profiling data.
Cloud Profiler User	View and query profiling data.

This completes this section on Cloud Profiler, where we looked at the supported profile types and learned how to use an interactive flame graph.

Binding SRE and Cloud Operations

Chapter 2, SRE Technical Practices – Deep Dive, introduced SRE technical practices such as SLAs, SLOs, SLIs, and Error Budgets. To summarize, this chapter established a relationship between these practices and tied them directly to the reliability of the service. To ensure that a service meets its SLAs, the service needs to be reliable. SRE recommends using SLOs to measure the reliability of the service. SLOs require SLIs to evaluate the service's reliability. If these SLIs are not met, then the SLOs will miss their targets. This will eventually burn the Error Budget, which is a measure that calculates the acceptable level of unavailability or unreliability. *Chapter 3, Understanding Monitoring and Alerting to Target Reliability*, introduced concepts related to monitoring, alerting, logging, and tracing and established how these are critical to tracking the reliability of the service. However, both these chapters were conceptual in nature.

This chapter's focus is Cloud Operations. So far, we've described how Google Cloud captures monitoring metrics, logging information, and traces and allows us to debug applications or services. Additionally, Cloud Operations has an option called SLO monitoring. This option allows you to define and track the SLO of a service. This option currently supports three service types for auto-ingestion: Anthos Service Mesh, Istio on GKE, and App Engine. However, this option also supports user-defined microservices. The next subsection deep dives into SLO monitoring.

SLO monitoring

Given that SLOs are measured using SLIs and SLOs are defined as quantifiable measures of service reliability that are measured over time, there are three specific steps in defining an SLO via SLO monitoring. These are as follows:

1. Setting a **SLI**
2. Defining SLI details
3. Setting a **SLO**

Let's look at these steps in more detail.

Setting an SLI

This is the first step and has two specific goals: choosing a metric as an SLI and selecting a method of evaluation for measuring the chosen metric.

Choosing a metric

SLO monitoring allows you to choose either **Availability** or **Latency** as an out-of-the-box SLI for a service that's been configured via Anthos Service Mesh, Istio on GKE, and App Engine. These options are not available for microservices on GKE that haven't been configured through the preceding options. These are also known as custom services. However, irrespective of how the service is configured, you have the option to choose **Other**. Here, the user can pick the metric of choice to track as the SLI.

Request-based or windows-based

There are two methods of evaluation to choose from that will affect how compliance against SLIs is measured. These are request-based and windows-based. The request-based option counts individual events and evaluates how a service performs over the compliance period, irrespective of how load is distributed. The windows-based option, on the other hand, measures performance in terms of time (good minutes versus bad minutes), irrespective of how load is distributed.

Defining SLI details

This is the second step and provides options for the user to choose a performance metric. The user can either use the predefined metrics in Cloud Monitoring or any user-defined metrics that can be created from logs (through logs-based metrics). Once a metric has been chosen, the performance criteria for the metric need to be defined. The performance criteria for metrics related to services on Anthos Service Mesh, Istio on GKE, and App Engine are predefined. However, for custom services, this needs to be manually defined by the user by using two of the three filter options – **Good**, **Bad**, and **Total**.

Setting an SLO

This is the third and final step and has two specific goals: setting the compliance period and setting the performance goal.

Compliance period

The compliance period option allows you to set a time period to evaluate the SLO. There are two possible choices:

- **Calendar**: Performance is measured from the start of the period, with a hard reset at the start of every new period. The available options for period length are Calendar day, Calendar week, Calendar fortnight, and Calendar month.

- **Rolling**: Performance is measured for a fixed time period; say, the last 10 days. The user can specify the fixed time period in days.

Now, let's look at setting the performance goal.

Performance goal

The performance goal indicates the goal that's been set as a ratio of *good service* to *demanded service* over the compliance period. This goal can be refined as more information is known about the system's behavior.

This completes our overview of SLO monitoring, which we can use to define an SLO to measure the reliability of a service. The next subsection provides a hands-on demonstration of how SLO monitoring can be configured against a GKE service (that we previously created in *Chapter 8, Understanding GKE Essentials to Deploy Containerized Applications*).

Hands-on lab – tracking service reliability using SLO monitoring

SLO monitoring allows us to link the SRE technical practices with the practical options available in Google Cloud. These help us monitor the reliability of the service and alert the on-call engineer if the service misses the reliability threshold.

This subsection is a hands-on lab that will show you how to use the SLO monitoring option from Cloud Monitoring. The SLO monitoring option tracks service reliability by defining an SLO. In this lab, we will use `hello-world-service` from the `my-first-cluster` GKE cluster, which was created as part of *Chapter 8, Understanding GKE Essentials to Deploy Containerized Applications*. This lab has three main goals:

- Defining an SLO for a service
- Creating an SLO burn rate alert policy
- Verifying SLO monitoring

Let's take a look at these goals in more detail.

Defining a SLO for a service

Follow these steps to define a SLO for `hello-world-service`:

1. Navigate to the **Services** UI under the **Monitoring** section of the GCP console. Select the **Define Service** action. Then, select the `hello-world-service` section from the `my-first-cluster` cluster, as shown in the following screenshot. Set the display name to `hello-world-service`. The system will create the service to be monitored and will navigate the user to the service overview dashboard:

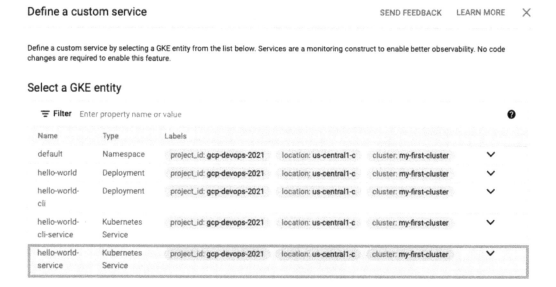

Figure 10.31 – Defining a custom service by selecting one

2. Select the **Create SLO** action to define an SLO. This action will open a pop-up window, as shown in the following screenshot. Note that, as discussed in *Chapter 2, SRE Technical Practices – Deep Dive*, an SLO requires an SLI. So, to define an SLO, we must first choose the SLI metric and then define it.

3. Given that the service being used for this lab is not part of Anthos Service Mesh, Istio on GKE, or App Engine, the only option available is to choose **Other**. Here, the user can configure a metric of choice to measure the performance of the service. In addition, set the method of evaluation to **Request-based**:

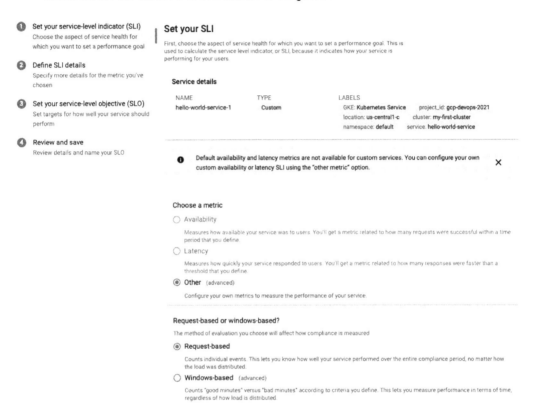

Figure 10.32 – Setting an SLI as part of SLO monitoring

4. To define the SLI's details, select a performance metric. In this case, we will select the `kubernetes.io/container/restart_count` metric. Set the filters to **Total** and **Bad**, as shown here:

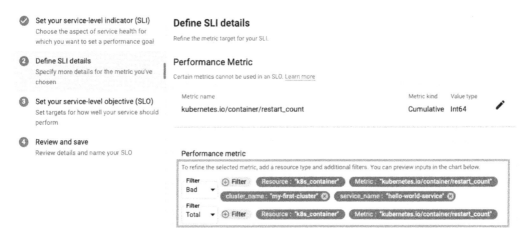

Figure 10.33 – Defining SLI details as part of SLO monitoring

5. Select a compliance period; that is, either **Calendar** or **Rolling**. For this specific lab, set **Period Type** to **Calendar** and **Period Length** to **Calendar day**. Additionally, set the performance goal to 90%, as shown here:

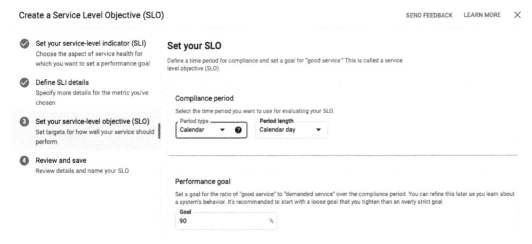

Figure 10.34 – Setting an SLO as part of SLO monitoring

6. Review the configuration and save it by providing an appropriate display name, such as `90% - Restart Count - Calendar Day`, as shown here:

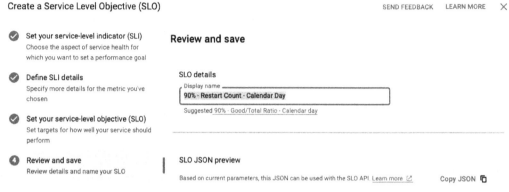

Figure 10.35 – Reviewing and saving the SLO as part of SLO monitoring

7. Once saved, the SLO – `90% - Restart Count - Calendar Day` – will be created under the `hello-world-service` service, as shown in the following screenshot. At the moment, the error budget is **100%** since none of the containers were restarted:

Figure 10.36 – SLO created for a service as part of SLO monitoring

With this, we've learned the steps we need to take to define an SLO for a service. In the next topic, we'll explore the steps we need to create an SLO burn rate alert policy.

Creating a SLO burn rate alert policy

The concept of alerting and notification channels from Cloud Monitoring (discussed earlier in this chapter) is used to create an alert. Before we look at the steps for this, let's recap on the critical jargon that was discussed in *Chapter 3, Understanding Monitoring and Alerting to Target Reliability*. We must configure these elements while defining an alert through Cloud Monitoring:

- **Lookback duration** refers to how far you must go back in time to retrieve monitoring data.

- **Fast-burn alert** refers to using shorter lookback durations that help with quickly detecting problems. However, this will lead to more frequent alerting and, potentially, false alarms.

- **Slow-burn alert** refers to using a longer lookback duration to ensure that a problem exists for a longer duration and avoids false alarms. However, the downside is that the alert is fired after a longer duration, even though the problem has a current negative impact on the service.

Follow these steps to set up an alert for when the error budget for the SLO drops beyond a certain burn rate within a specified period of time:

1. Click the **Create alerting policy** button, as shown in the preceding screenshot, to create an SLO alerting policy. The SLO alerting policy tracks the error budget based on the configured burn rate, as shown in the following screenshot. Set **Lookback duration** to 1 minute(s) and **Burn rate threshold** to 10 minute(s). The following is our configuration for a fast-burn alert:

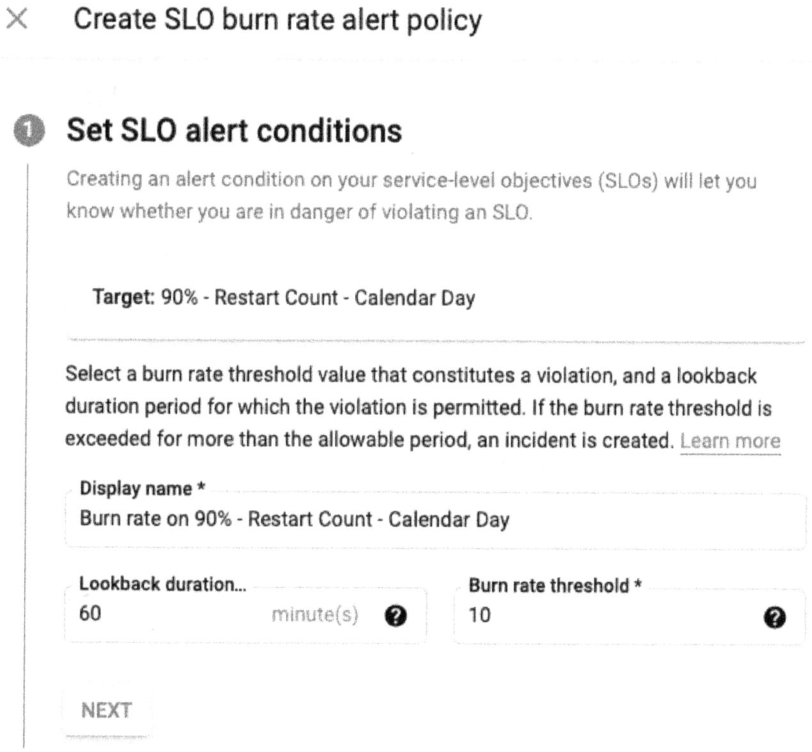

×   Create SLO burn rate alert policy

① Set SLO alert conditions

Creating an alert condition on your service-level objectives (SLOs) will let you know whether you are in danger of violating an SLO.

Target: 90% - Restart Count - Calendar Day

Select a burn rate threshold value that constitutes a violation, and a lookback duration period for which the violation is permitted. If the burn rate threshold is exceeded for more than the allowable period, an incident is created. Learn more

Display name *
Burn rate on 90% - Restart Count - Calendar Day

Lookback duration...		Burn rate threshold *	
60	minute(s) ❷	10	❷

NEXT

Figure 10.37 – Setting an SLO alert condition

2. Select a notification channel of choice (that has already been pre-configured). In this case, select an email notification channel, as shown here:

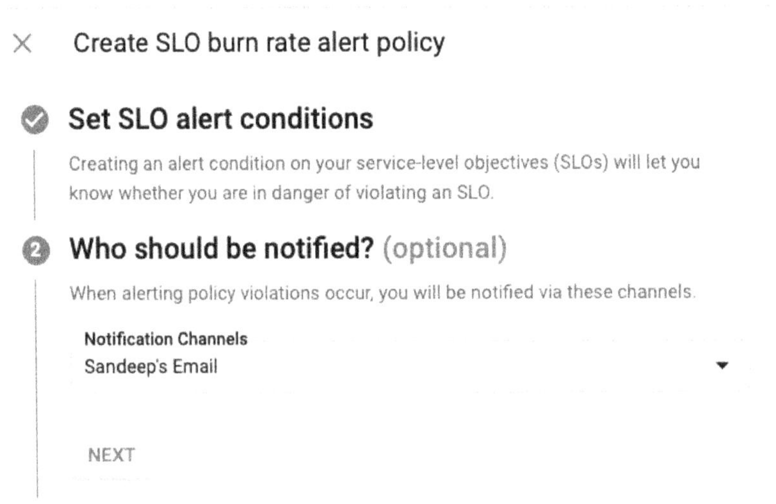

✕   Create SLO burn rate alert policy

☑ **Set SLO alert conditions**

Creating an alert condition on your service-level objectives (SLOs) will let you know whether you are in danger of violating an SLO.

② **Who should be notified?** (optional)

When alerting policy violations occur, you will be notified via these channels.

Notification Channels
Sandeep's Email   ▼

NEXT

Figure 10.38 – Selecting a notification channel to send an alert

Now, create the SLO burn rate alert policy. Optionally, add documentation that references the alert in terms of what the on-job SRE engineer should check or do.

3. Once the alert has been configured, the SLO status will look as follows, where **Error Budget** is currently at 100% and none of the alerts are firing:

Figure 10.39 – Showing the complete setup for SLO monitoring with its alert and initial error budget

With that, we've created an SLO burn rate alert policy. Now, let's verify SLO monitoring by performing a test.

Verifying SLO monitoring

In the previous two subsections of this hands-on lab on SLO monitoring, we created an SLO for a service (that was previously created in *Chapter 8*, *Understanding GKE Essentials to Deploy Containerized Applications*) and then created an SLO burn rate alert policy. This section will show you how to test the configuration and verify if our SLO monitoring option verifies the health of our service; that is, `hello-world-service`:

1. Given that we previously selected the performance metric while defining our SLO as `kubernetes.io/container/restart_count`, let's restart the container and see if the error budget changes and, subsequently, if the alert gets fired. Use the following command to restart the container after connecting to the cluster. Replace `pod-name` and `container-name` accordingly. `pod-name` can be found via the service, while `container-name` can be found via `pod-name`:

```
kubectl exec -it <pod-name> -c <container-name> -- /bin/
sh -c "kill 1"
```

2. Once the command has been executed, the container inside the pod, with respect to `hello-world-service`, will restart. This means that the SLI that's been defined will not be met and, subsequently, the SLO will not be met. As a result, the error budget will be consumed. If the error budget is consumed by more than the burn rate that's been defined – which was 10 under 1 minute – then an alert will also be fired. The following screenshot shows the updated status of the SLO for `hello-world-service`. The status of the SLO has now been updated to **Unhealthy**:

Figure 10.40 – Displaying the service as Unhealthy, alerts firing, and the reduced error budget

3. The alert triggers a notification that will be sent to the configured email, as shown in the following screenshot:

 Alert firing

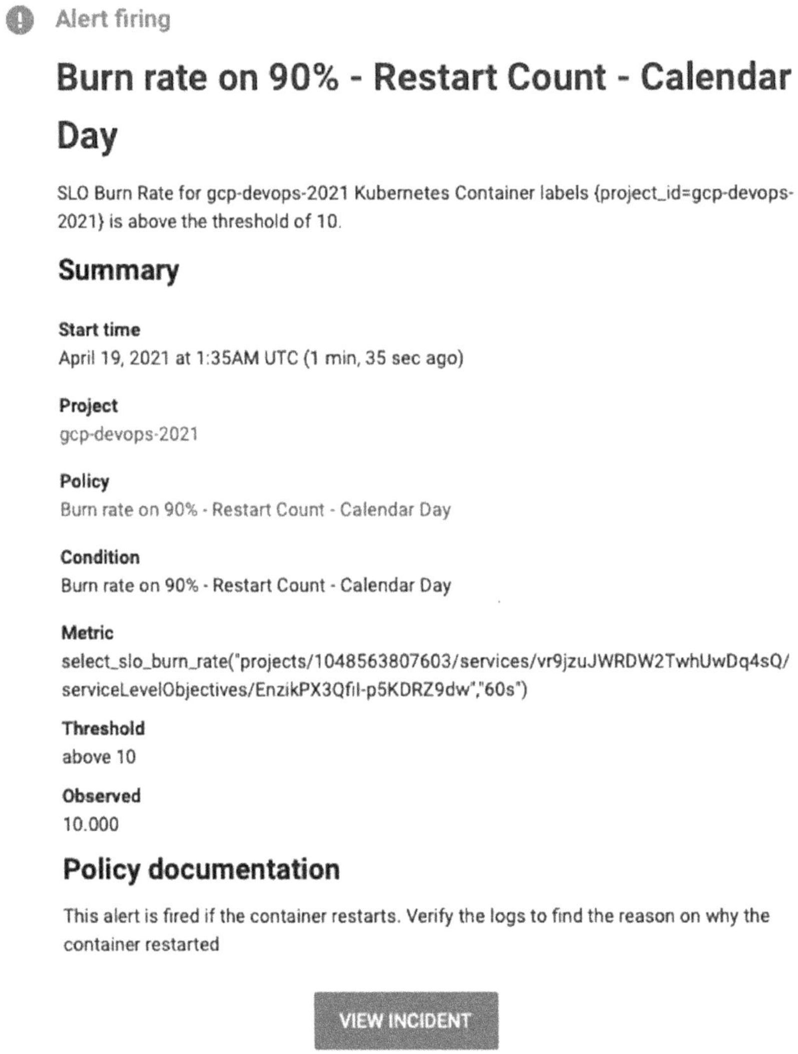

Burn rate on 90% - Restart Count - Calendar Day

SLO Burn Rate for gcp-devops-2021 Kubernetes Container labels {project_id=gcp-devops-2021} is above the threshold of 10.

Summary

Start time
April 19, 2021 at 1:35AM UTC (1 min, 35 sec ago)

Project
gcp-devops-2021

Policy
Burn rate on 90% - Restart Count - Calendar Day

Condition
Burn rate on 90% - Restart Count - Calendar Day

Metric
select_slo_burn_rate("projects/1048563807603/services/vr9jzuJWRDW2TwhUwDq4sQ/serviceLevelObjectives/EnzikPX3Qfil-p5KDRZ9dw","60s")

Threshold
above 10

Observed
10.000

Policy documentation

This alert is fired if the container restarts. Verify the logs to find the reason on why the container restarted

VIEW INCIDENT

Figure 10.41 – Alert notification set to the configured email address

This completes our detailed hands-on lab related to SLO monitoring, where we linked the SRE technical practices to options available in Google Cloud Operations to monitor and alert users about the reliability of the service. This also completes this chapter on Cloud Operations.

Summary

In this chapter, we discussed the suite of tools that are part of Cloud Operations. Cloud Operations is critical for forming the feedback loop of the CI/CD process and is fundamental to establishing observability on GCP. Observability is key to ensuring that an SRE's technical practices – specifically, SLIs, SLOs, SLAs, and Error Budgets – are not violated. This is achieved by gathering logs, metrics, and traces from multiple sources and by visualizing this information on dashboards. This information is used to establish performance and reliability indicators. These indicators can then be tracked with configurable alerts. These alerts trigger when there is a potential violation, and the alerts will be notified on the configurable notification channels. Cloud Operations also offers services that allow us to debug the application, without slowing down, and capture trace information. The end goal is to ensure that the service is reliable. We concluded this chapter by providing a hands-on lab on SLO monitoring, a feature from Google Cloud that tracks the reliability of the service by bringing together Cloud Operations and SRE technical practices.

This was the last chapter of this book. The next section provides insights into preparing to become a Professional Cloud DevOps Engineer, along with a summary on a few topics that might show up in the exam but were not covered in the last 10 chapters. We have also provided a mock exam, which will be useful as a preparation resource.

Points to remember

The following are some important points to remember:

- Cloud Monitoring is a GCP service that collects metrics, events, and metadata from multi-cloud and hybrid infrastructures in real time.

- A workspace provides a *single pane of glass* related to GCP resources.

- A workspace can monitor resources from multiple monitored projects.

- A monitored project, however, can only be associated with a single workspace.

- Dashboards provide a graphical representation of key signal data, called metrics, in a manner that is suitable for end users or the operations team.

- Metrics represent numerical measurements of resource usage that can be observed and collected across the system at regular time intervals.

- MQL can be used to create a chart with a text-based interface and uses an expressive query language to execute complex queries against time series data.

- Uptime checks test the availability of an external facing service within a specific timeout interval.

- Connection errors, 40x client errors, and not configuring firewall rules are potential reasons for uptime check failures.

- Alerting is the process of processing the alerting rules that track the SLOs and notify or perform certain actions when the rules are violated.

- The alignment period is a lookback interval from a particular point in time.

- The Monitoring agent is based on the `collectd` daemon and is used to collect system statistics from various sources, including OSes, applications, logs, and external devices.

- Cloud Logging is a GCP service that allows you to store, search, analyze, monitor, and alert users about logging data and events from applications.

- Policy denied logs are specific to logs that are captured when access is denied by a Google Cloud service to either a user or service account.

- Logs-based metrics are metrics that are created based on the content of the log entries and can be extracted from both included and excluded logs.

- VPC Flow Logs capture real-time network activity (incoming/outgoing) against VPC resources on an enabled subnet.

- Firewall logs capture the effects of a specific firewall rule in terms of the traffic that's allowed or denied by that firewall rule.

- The Logging agent is based on `fluentd` and captures additional VM logs such as operating system (OS) logs and logs from third-party applications.

- The Monitoring and Logging agents can both be installed on unmanaged GCE VMs.

- GKE has built-in support for logging and can be enabled for new or existing clusters via *Cloud Operations for GKE*.

- Cloud Debugger inspects the state of a running application in real time.

- Snapshots capture local variables and the call stack at a specific location in the application's source code.

- A snapshot is only taken once, and the user needs to manually retake it if needed.

- Debug logpoints can inject a log into a running application without stopping, editing, or restarting.

- Logpoints can be created even if direct access to the source code is not available.

- Logpoints become inactive after 24 hours and are automatically deleted after 30 days.

- Cloud Trace is a collection of spans. A span is an object that wraps latency-specific metrics. Cloud Trace is a distributed tracing system.

- Cloud Trace's language-specific SDKs are available for Java, Node.js, Ruby, and Go.

- Cloud Profiler provides low-impact continuous profiling to help us understand the performance of a production system.

- The programming languages that are supported by Cloud Profiler include Java, Go, Node.js, and Python. Profile data is retained for 30 days by default.

Further reading

For more information on GCP's approach toward DevOps, please read the following articles:

- **Cloud Operations**: `https://cloud.google.com/products/operations`
- **Cloud Monitoring**: `https://cloud.google.com/monitoring`
- **Cloud Logging**: `https://cloud.google.com/logging`
- **Cloud Debugger**: `https://cloud.google.com/debugger`
- **Cloud Trace**: `https://cloud.google.com/trace`
- **Cloud Profiler**: `https://cloud.google.com/profiler`

Practice test

Answer the following questions:

1. A user has performed administrative actions that modify the configuration or metadata of resources. Which of the following is the most appropriate option to quickly get to the logs related to administrative actions?

   a) Go to Error Reporting and view the administrative activity logs.

   b) Go to Cloud Logging and view the administrative activity logs.

   c) Go to Cloud Monitoring and view the administrative activity logs.

   d) Go to the Activity tab on the Cloud Console and view the administrative activity logs.

2. The default retention period for data access audit logs is _____.

   a) 7 days

   b) 30 days

   c) 400 days

   d) Unlimited

3. Select the most appropriate option for monitoring multiple GCP projects with resources through a single workspace.

   a) Cannot monitor multiple GCP projects through a single workspace.

   b) Configure a separate project as a host project for a Cloud Monitoring workspace. Configure metrics and logs from each project to the host project via Pub/Sub.

   c) Configure a separate project as a host project for a Cloud Monitoring workspace. Use this host project to manage all other projects.

   d) Configure a separate project as a host project for a Cloud Monitoring workspace. Configure the metrics and logs from each project for the host project via Cloud Storage.

4. _____ logs record operations of instances that have been reset for Google Compute Engine.

   a) Admin activity

   b) System event

   c) Data access

   d) Access transparency

5. The maximum size of a log entry is _____.

   a) 64 KB

   b) 128 KB

   c) 256 KB

   d) 512 KB

6. The default retention period for access transparency logs is _____.

    a) 7 days

    b) 30 days

    c) 400 days

    d) Unlimited

7. _____ logs are specific to actions that are performed by Google personnel when accessing user's/customer's content.

    a) Admin activity

    b) System event

    c) Data access

    d) Access transparency

8. The SRE team supports multiple production workloads in GCP. The SRE team wants to manage issues better by sending error reports and stack traces to a centralized service. Which of the following is best suited for accomplishing this goal?

    a) Cloud Error Logging

    b) Cloud Error Reporting

    c) Cloud Tracing

    d) Cloud Profiling

9. _____ logs record the operations that are performed when assigning/ unassigning IAM roles.

    a) Admin activity

    b) System event

    c) Data access

    d) Access transparency

10. _____ logs analyze the network logs of an application.

    a) VPC flow

    b) Firewall

    c) Audit

    d) Activity

11. Select the option that represents the right characteristics for log entry from Cloud Logging:

   a) Timestamp

   b) Log name

   c) Resource tied to the log entry

   d) All of the above

12. Select two actions where the user will want to send a subset of logs for big data analysis:

   a) Create a sink in Cloud Logging that identifies the subset of logs to send.

   b) Export logs to Cloud Storage.

   c) Export logs to BigQuery.

   d) Export logs to Pub/Sub.

13. The default retention period for admin activity logs is _____.

   a) 7 days

   b) 30 days

   c) 400 days

   d) Unlimited

14. Which of the following represents the right sequence of steps to export logs?

   a) Choose destination, create sink, create filter

   b) Create sink, create filter, choose destination

   c) Create sink, choose destination, create filter

   d) Choose destination, create filter, create Sink

15. _____ logs will record how resources are created for Google Compute Engine:

   a) Admin activity

   b) System event

   c) Data access

   d) Access transparency

16. Select the option that governs access to logs from Cloud Logging for a given user:

    a) Service accounts

    b) Cloud IAM roles

    c) Both (a) and (b)

    d) None of the above

17. Select the role that allows us to manage IAM roles for a Monitoring workspace:

    a) Monitoring Viewer

    b) Monitoring Editor

    c) Monitoring Admin

    d) Monitoring Metric Writer

18. Select the Cloud Monitoring widget that represents metrics with a distribution value:

    a) Line charts

    b) Heatmap charts

    c) Gauges

    d) Scorecards

19. To perform uptime checks, what is the minimum number of active locations that need to be selected as geographic regions?

    a) Two

    b) Three

    c) Four

    d) Five

20. The Monitoring agent is based on _____, while the Logging agent is based on _____.

    a) `fluentd, collectd`

    b) `google-collectd, google-fluentd`

    c) `collectd, fluentd`

    d) `google-fluentd, google-collectd`

21. Which of the following is not a valid classification type for data access logs?

    a) Admin read

    b) Admin write

    c) Data read

    d) Data write

22. Select the role that allows us to view data access and access transparency logs:

    a) Logs Viewer

    b) Private Logs Viewer

    c) Project Viewer

    d) Project Editor

23. The default retention period for firewall logs is _____.

    a) 7 days

    b) 30 days

    c) 400 days

    d) Unlimited

24. Every VPC has a set of hidden, implied, pre-configured rules with the lowest priority. Select two valid pre-configured rules:

    a) `allow all ingress`

    b) `deny all ingress`

    c) `allow all egress`

    d) `deny all egress`

25. The default retention period for system event audit logs is _____.

    a) 7 days

    b) 30 days

    c) 400 days

    d) Unlimited

Answers

1. (d): Go to the Activity tab on the Cloud Console and view the administrative activity logs.

2. (b): 30 days.

3. (c): Configure a separate project as a `host` project for a Cloud Monitoring workspace. Use this host project to manage all other projects.

4. (b): System event.

5. (c): 256 KB.

6. (c): 400 days.

7. (d): Access transparency.

8. (b): Cloud Error Reporting.

9. (a): Admin activity.

10. (a): VPC flow.

11. (d): All of the above.

12. (a) and (c).

13. (c): 400 days.

14. (b): Create sink, create filter, choose destination.

15. (a): Admin activity.

16. (b): Cloud IAM roles.

17. (c): Monitoring Admin.

18. (b): Heatmap chart.

19. (b): Three.

20. (c): `collectd, fluentd`.

21. (b): Admin write; this is not a valid classification for data access logs.

22. (b): Private Logs Viewer.

23. (b): 30 days.

24. (b) and (c): `deny all ingress` and `allow all egress`.

25. (c): 400 days.

Appendix
Getting Ready for Professional Cloud DevOps Engineer Certification

This book is a practical guide to learning about and understanding **site reliability engineering** or **SRE**, which is a prescriptive way of implementing DevOps. The book also provides deep insights into the Google Cloud services that are critical to implementing DevOps on Google Cloud Platform.

Additionally, the book also helps in preparing for the Professional Cloud DevOps Engineer Certification exam. A professional Cloud DevOps engineer is responsible for efficient development operations and balancing service reliability and delivery speed. They are skilled at using Google Cloud Platform to build software delivery pipelines, deploy and monitor services, and manage and learn from incidents. The official exam guide can be found at `https://cloud.google.com/certification/guides/cloud-devops-engineer`. To register for the certification exam, go to `https://cloud.google.com/certification/register`.

At a high level, the certification is centered around SRE, **Google Kubernetes Engine** (**GKE**), and Google Cloud's operations suite. These topics probably make up more than 80% of the certification exam. The chapters on SRE, GKE, and Google Cloud's operations suite extensively cover key concepts assessed in the Professional Cloud DevOps Engineer Certification exam. The *Points to remember* section and the practice test toward the end of this chapter should help you revise the critical concepts from this chapter.

In addition, three additional topics that might come up on the certification exam will be covered at a high level. Some of these topics overlap with other certification exams from Google, such as Professional Cloud Architect Certification or Professional Cloud Developer Certification.

The following are the topics that will be summarized. Please note that these topics are not completely elaborated on here but rather only introduced. It is recommended to refer to the specific documentation on these topics for in-depth information:

- Cloud Deployment Manager
- Cloud Tasks
- Spinnaker

Cloud Deployment Manager

Infrastructure as Code (**IaC**) is the process of managing and provisioning infrastructure through code instead of manually creating the required resources. **Cloud Deployment Manager** is a Google Cloud service that provides IaC. Cloud Deployment Manager can create a set of Google Cloud resources and facilitates managing these resources as a unit otherwise called a deployment. For example, it is possible to create a **Virtual Private Cloud** (**VPC**) using declarative code through a configuration file rather than manually creating it through the console. The following are some critical properties of Cloud Deployment Manager:

- Can create multiple resources in parallel, such as multiple VMs
- Can provide input variables to create a resource with specific user-defined values as required
- Can get the return value of a newly created resource, such as the instance ID of a newly created Google Compute Engine instance
- Can create dependencies where one resource definition can reference another resource and one resource can be created after creating another resource (using `dependsOn`)

Cloud Deployment Manager allows specifying all the resources required for an application through a configuration file. This is the first step for implementing Cloud Deployment Manager. This configuration file is written in a declarative format using YAML.

Each configuration file can be used to define one or more resources. Each resource section consists of three main components: the name of the resource, the resource type, and the resource properties. The resource properties that need to be used are in most cases specific to the resource type. The resources specified in the configuration file are created by making API calls (which could introduce a slight risk as certain APIs could be deprecated in the future). The configuration file can either be completely spelled out or Cloud Deployment Manager allows the usage of templates (which are preferred for creating similar types of resources).

A configuration can contain templates that refer to parts of the configuration file that is abstracted to individual building blocks. A template file is written in either Python or Jinja2. Python templates are more powerful and provide the option to programmatically create or manage templates. Jinja2 is a simpler but less powerful templating language that uses the same syntax as YAML. A preview mode (using the `--preview` flag) can be used to verify the potential operations on the resources before they are applied. For more information on Cloud Deployment Manager, refer to `https://cloud.google.com/ deployment-manager`.

Cloud Tasks

Cloud Tasks is a fully managed service from Google Cloud that allows you to separate out pieces of work that could be performed independently and asynchronously outside of a user or a service-to-service request. An independent piece of work is referred to as a task. Cloud Tasks is essentially used when an application accepts inputs from users and needs to initiate background tasks accordingly to perform automated asynchronous execution.

The following is a summary of the critical features of Cloud Tasks:

- Cloud Tasks is aimed at explicit invocation, where the publisher retains full control of execution.

- Cloud Tasks is most appropriate where the task producer can have control over the execution.

The core difference between Cloud Tasks and Pub/Sub is the notion of explicit versus implicit invocation. As mentioned, Cloud Tasks is aimed at explicit invocation. In contrast, Pub/Sub supports implicit invocation, where a publisher implicitly causes the subscriber to execute by publishing an event. For more in-depth information on Cloud Tasks, refer to `https://cloud.google.com/tasks`.

Spinnaker

Spinnaker is an open source, multi-cloud continuous delivery platform that was initially developed by Netflix and later extended by Google. Spinnaker is not an official Google Cloud service. Spinnaker is not a natively integrated service, and no such native service exists yet from Google Cloud Platform for continuous deployment. Spinnaker is extensively recommended by Google to implement CI/CD pipelines on Google Cloud Platform. Spinnaker helps to release software changes at high velocity and with confidence. Spinnaker is composed of several independent microservices. Spinnaker creates a Cloud Storage bucket and uses a Redis database to maintain its assets. Spinnaker also creates a Pub/Sub topic.

Spinnaker can be considered an application management tool where it is possible to view and manage GKE components, including workload resources, services, and load balancers. Spinnaker can support multiple deployment models, including rolling update, blue/green, or canary.

Spinnaker puts everything in an auto-pilot mode where there is no need for manual intervention, or in other words, no need to manually execute `kubectl` commands. Spinnaker can create and manage YAML files and can perform automated deployments or even create/execute YAML files for load balancers as a service. The only manual step involved is to provide approval to complete deployment.

The following figure shows a high-level summary in terms of interactions specifically when Spinnaker is used to deploy to a GKE cluster:

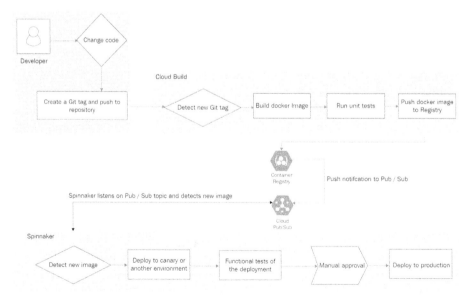

Illustration of interactions when Spinnaker is used to deploy to GKE

The interactions in the preceding figure are as follows:

1. The developer changes the code by creating a Git tag and pushes to the cloud source repository.

2. The cloud source repository is configured to detect the new Git tag. This triggers Cloud Build to execute the build process as per the provided specification. This might include the running of unit tests. Once complete, a build artifact such as a Docker image is created.

3. The created build artifact is stored either in Container Registry or Artifact Registry and once available, a message is sent to Cloud Pub/Sub if configured.

4. Spinnaker, installed on GKE, will be listening to a Pub/Sub topic for the newly created image.

5. Once the image is available, Spinnaker deploys the new container to a QA/staging environment, preferably through canary deployment so that only a small set of users are impacted.

6. Functional tests are run on the canary environment if available to verify the quality of the deployment. If the deployment is as per the expected standards in terms of quality, then after manual approval, the image is then deployed to the production environment.

It is important to note that either while performing a blue/green deployment or a canary deployment, Spinnaker can update the ReplicaSet in place. When deploying a ReplicaSet with a Deployment, and the Deployment doesn't exist, Spinnaker first creates the ReplicaSet with 0 replicas and then creates the Deployment, which will resize the ReplicaSet. When the Deployment does exist, it does the same but edits the Deployment in place rather than creating it.

For more information on installing and managing Spinnaker on Google Cloud Platform, refer to https://cloud.google.com/docs/ci-cd/spinnaker/spinnaker-for-gcp.

This concludes our summary of additional topics such as Cloud Deployment Manager, Cloud Tasks, and Spinnaker.

You have reached the end of this book. Test your knowledge by attempting the mock tests. This book includes 2 mock tests of 50 questions each. These tests can be used as a reference. All the best if you are taking the certification exam!

Mock Exam 1

Test Duration: 2 hours

Total Number of Questions: 50

Answer the following questions:

1. Which of the following is an appropriate deployment solution with minimal downtime and infrastructure needs?

   a) Recreate

   b) Rolling update

   c) Canary deployment

   d) Blue/green deployment

2. Cloud Pub/Sub and Cloud Tasks are two possible services in Google Cloud that provide the ability to asynchronously integrate with message services. Select two options that are the most appropriate for Cloud Tasks:

   a) Cloud Tasks is aimed at explicit invocation where the publisher retains full control over execution.

   b) Cloud Tasks is most appropriate where the task producer can have control over execution.

   c) Cloud Tasks is aimed at implicit invocation where the publisher retains full control over execution.

   d) Cloud Tasks is most appropriate where the task consumer can have control over execution.

3. _____ logs will record operations of instance reset for Compute Engine.

   a) Admin Activity

   b) System Event

   c) Data Access

   d) Access Transparency

4. Select the appropriate service type where a service gets an internal IP address:

   a) `ClusterIP`

   b) `NodePort`

   c) `LoadBalancer`

   d) All of the above

5. Select the role that allows the writing of monitoring data to a workspace but doesn't permit viewing via a console:

   a) Monitoring Viewer

   b) Monitoring Metric Writer

   c) Monitoring Admin

   d) Monitoring Editor

6. For a service, if the SLA is 99.0%, the SLO is 99.5%, and the SLI is 99.6%, then the error budget is _____.

   a) 1%

   b) 0.4%

   c) 0.5%

   d) None of the above

7. _____ is a feature of Cloud Build where intermediate container image layers are directly written to Google's container registry without an explicit push step.

   a) Elastic cache

   b) Kaniko cache

   c) Redis cache

   d) (a) and (c)

8. Who of the following coordinates the effort of the response team to address an active incident?

   a) **Incident Commander (IC)**

   b) **Communications Lead (CL)**

   c) **Operations Lead (OL)**

   d) **Planning Lead (PL)**

9. Select the command that allows Docker to use `gcloud` to authenticate requests to Container Registry?

   a) `gcloud auth configure docker`

   b) `gcloud auth configure-docker`

   c) `gcloud auth docker-configure`

   d) `gcloud auth docker configure`

10. Select the phase as per the SRE engagement model where an SRE engineer is less engaged in comparison with the other phases:

    a) Architecture and design

    b) Active development

    c) Limited availability

    d) General availability

11. Your application is deployed in GKE and utilizes secrets to protect sensitive information. Which of the following is the recommended approach?

    a) Sensitive information should not be stored in the application.

    b) Sensitive information should be passed as environment variables.

    c) Sensitive information should be stored in the cloud's DLP service.

    d) Sensitive information should be stored in Google Secrets Manager.

12. A logging agent is installed on Google Compute Engine. However, logs are not sent to Cloud Logging. What should be the next logical step to troubleshoot this issue?

    a) Re-install the logging agent.

    b) Restart Compute Engine.

    c) Configure data access logs as part of the logging agent configuration.

    d) Verify whether the service account tied to Compute Engine has the required role to send to Cloud Logging.

13. The SLI is calculated as *Good Events / Valid Events * 100*. Which of the following response counts are discarded while calculating valid events?

    a) 2xx status count

    b) 3xx status count

    c) 4xx status count

    d) 5xx status count

14. Deployment rollout is triggered only if the deployment's pod template is changed. Select the option that represents the section of the deployment that should be changed:

    a) `.spec`

    b) `.spec.template`

    c) `.template.spec`

    d) `.template`

15. Which of the following is not true as per SRE best practices?

    a) There are consequences for missing an SLA.

    b) There are no consequences for missing an SLO.

    c) A missing SLO represents an unhappy user.

    d) A missing SLO results in burning the error budget.

16. Select the log type that represents log entries to perform read-only operations without modifying any data:

    a) Admin Activity

    b) System Event

    c) Data Access

    d) Access Transparency

17. Which of the following is not a recommended best practice during the incident management process?

    a) Develop and document procedures.

    b) The Incident Commander should sign off on all decisions during incident management.

c) Prioritize the damage and restore the service.

d) Trust team members in specified roles.

18. _____ represents an exact measure of a service's behavior.

a) SLI

b) SLO

c) SLA

d) Error budget

19. An application in App Engine Standard is generating a large number of logs. Select the appropriate option that quickly provides access to errors generated by the application?

a) Go to Cloud Console, the Activity tab, and view the error logs.

b) Go to Cloud Logging, filter the logs, and view the error logs.

c) Go to Cloud Error Reporting and view the error logs aggregated by occurrences.

d) Go to Cloud Monitoring, filter the logs, and view the error logs.

20. _____ logs will record operations in terms of listing resources for Compute Engine.

a) Admin Activity

b) System Event

c) Data Access

d) Access Transparency

21. You are planning to adopt SRE practices and want to set an SLO for your application. How do you get started?

a) Set the SLO to 100%.

b) Set the SLO based on collecting historical data during the pre-GA period.

c) Set the SLO based on an internal survey of all stakeholders.

d) Set the SLO slightly below the SLA.

22. Select the appropriate option that describes how to delete admin activity logs:

    a) Go to the Activity tab in Cloud Console. Filter the logs to delete and select the delete action.

    b) Go to the Activity tab in Cloud Logging. Filter the logs to delete and select the delete action.

    c) Admin activity logs cannot be deleted.

    d) Go to the Activity tab in Cloud Monitoring. Filter the logs to delete and select the delete action.

23. Which of the following is the preferred method if Cloud Build needs to communicate with a separate CD tool regarding the availability of a new artifact, which could be deployed by the CD tool?

    a) Cloud Build makes an entry in a NoSQL database. The CD tool looks for an event in the database.

    b) Cloud Build makes an entry in a cloud storage bucket. The CD tool looks for an event in the bucket.

    c) Cloud Build publishes a message to a pub/sub topic. The CD tool is subscribed to the topic.

    d) (a) and (b).

24. Select all the possible characteristics of Toil (you may choose more than one):

    a) It is repetitive.

    b) It provides enduring value.

    c) It grows linearly with the service.

    d) Tactical.

25. _____ is a process/service that digitally checks each component of the software supply chain, ensuring the quality and integrity of software before the application is deployed.

    a) Container registry

    b) Cloud storage

    c) Container analysis

    d) Binary authorization

26. How many times does Cloud Debugger allow you to take a snapshot for application data for the same location in the code?

    a) None

    b) Twice

    c) Once

    d) Thrice

27. Select the option that is true in the case of audit security logs.

    a) The retention period is 30 days and is configurable.

    b) The retention period is 400 days and is configurable.

    c) The retention period is 30 days and is not configurable.

    d) The retention period is 400 days and is not configurable.

28. _____ API enabled metadata storage regarding software artifacts and is used during the binary authorization process

    a) Container scanning

    b) Container metadata

    c) Container attestation

    d) Container analysis

29. You have created a GKE cluster with binary authorization enabled. You are trying to deploy an application, but you encountered an error – *Denied by Attestor*. Select the next appropriate step.

    a) Update the cluster and turn off binary authorization. Deploy the application again.

    b) Create cryptographic keys and submit to binary authorization.

    c) Create an attestation and submit to binary authorization.

    d) Create an exemption for the container image from binary authorization.

30. Select the account to use in code when one Google cloud service interacts with another Google cloud service:

    a) User account

    b) Service account

    c) Both (a) and (b)

    d) None of the above

31. Your application has an API that serves responses in multiple formats, such as JSON, CSV, and XML. You want to know the most requested format for further analysis. Which of the following is the most appropriate approach?

    a) Create a log filter and export the logs to cloud storage through a log sink. Query the logs in cloud storage.

    b) Create a log filter, create a custom metric using regex, and extract each requested format into a separate metric. Monitor the count of metrics and create alerts as needed.

    c) Create a log filter and export the logs to BigQuery through a log sink. Query the logs in BigQuery.

    d) Create a log filter, create a custom metric using regex, and extract the requested format to a label. Monitor the metric to view the count of requests grouped by label.

32. Select the incident severity classification that has the following characteristics: an impact on, or inconvenience to, internal users, but external users might not notice.

    a) Negligible

    b) Minor

    c) Major

    d) Detrimental

33. Select the log type that represents the log entries for system maintenance operations on Compute Engine resources?

    a) Admin Activity

    b) System Event

    c) Data Access

    d) Access Transparency

34. Which of the following is a suitable SLI for big data systems (select two)?

    a) Availability

    b) Throughput

    c) Quality

    d) End-to-end latency

35. You need to deploy two applications represented by two deployments – Alpha and Beta. Both the applications need to communicate with each other, but none of the applications should be reachable from the outside. You decided to create a service. Select the appropriate service type:

    a) `Ingress`

    b) `NodePort`

    c) `ClusterIP`

    d) `LoadBalancer`

36. There is an expectation to auditing network traffic inside a specific VPC. Which of the following is the most appropriate option?

    a) Enable VPC firewall-level logs

    b) Enable VPC network logs

    c) Enable VPC flow logs

    d) Enable VPC audit logs

37. In RBAC, `persistentvolume` and `certificatesigningrequests` nodes are scoped as:

    a) Cluster scoped

    b) Namespace scoped

    c) Both

    d) None

38. Select the resource that bundles application code and dependencies into a single unit, leading to the abstraction of the application from the infrastructure.

    a) Docker

    b) Microservice

    c) Containers

    d) Virtual machines

39. VPC in Google Cloud is a _____ resource.

    a) Regional

    b) Zonal

    c) Global

    d) Multi-regional

40. An application deployed in a GKE cluster is having an intermittent problem. Every time a problem happens, the application is emitting a very specific set of logs. Your intention is to debug the application or see the state of the application as soon as the situation happens. Which of the following is the best possible approach?

   a) Cloud monitoring: Set up a custom metric based on the specific log pattern and create an alert from the metric when the number of lines for the pattern exceeds a defined threshold.

   b) Cloud logging: Create a log filter on a specific log pattern, export logs to BigQuery by creating a log sink, and create alerts by monitoring data in BigQuery based on specific patterns.

   c) Cloud error reporting: Set up a custom metric based on the specific log pattern and create an alert from the metric when the number of lines for the pattern exceeds a defined threshold.

   d) Cloud logging: Set up a custom metric based on the specific log pattern and create an alert from the metric when the number of lines for the pattern exceeds a defined threshold.

41. Select the option that is not in accordance with Google's recommendations vis-à-vis cloud-native development:

   a) Use microservice architectural patterns.

   b) Design components to be stateful wherever possible.

   c) Design for automation.

   d) Build everything as containers.

42. You need to deploy a container image to a GKE cluster. A cluster is not yet created. Which of the following options presents the most appropriate sequence of steps?

   a) Create a GKE cluster using the kubectl command, create a deployment specification using the container image, and use kubectl to create the deployment.

   b) Create a GKE cluster using the gcloud command, create a deployment specification using the container image, and use kubectl to create the deployment.

   c) Create a GKE cluster using the kubectl command, create a deployment specification using the container image, and use gcloud to create the deployment.

   d) Create a GKE cluster using the gcloud command, create a deployment specification using the container image, and use gcloud to create the deployment.

43. The smallest unit of virtualized hardware is known as a _____.

    a) Pod

    b) Node

    c) Job

    d) Container

44. The goal is to create a GKE cluster and use a third-party monitoring application. Which of the following is the most appropriate way to implement this?

    a) Deploy the monitoring application as a `ReplicaSet` object.

    b) Deploy the monitoring application as a `DaemonSet` object.

    c) Deploy the monitoring application as a `StatefulSet` object.

    d) Deploy the monitoring application as a `LoadBalancer` service.

45. Select the controller used by Kubernetes Engine for stateful applications.

    a) Stateless controller

    b) StatefulSet controller

    c) DaemonSet controller

    d) Replication controller

46. There is a need to retain logs for at least 3 years for analytical purposes. Which of the following is the most preferred approach?

    a) Cloud logging: Logs are available for a configurable amount of time.

    b) Cloud logging: Create a log sink and export it to cloud storage.

    c) Cloud logging: Create a log sink and export it to Big Query.

    d) Cloud logging: Create a log sink and export it to Pub/Sub.

47. _____ is a precise numerical target for system availability.

    a) SLI

    b) SLO

    c) SLA

    d) Error budget

48. Sam needs to view the access transparency logs. Select the role that is most appropriate for this purpose?

    a) Project Editor

    b) Project Viewer

    c) Logs Viewer

    d) None of the above

49. Who of the following is responsible for the communications portion of the response?

    a) **Incident Commander (IC)**

    b) **Communications Lead (CL)**

    c) **Operations Lead (OL)**

    d) **Planning Lead (PL)**

50. Company A wants to set up alerting policies based on the burn rate (with respect to error budgets). Which of the following is the best option as per SRE recommended practices?

    a) Create an alerting policy with a shorter lookback period.

    b) Create an alerting policy that considers a shorter lookback period and a longer lookback period.

    c) Create an alerting policy with a longer lookback period.

    d) Create an alerting policy whenever a defined condition is violated instantly.

Answers

1. (b) – Rolling update
2. (a) and (b)
3. (b) – System Event
4. (d) – All of the above. Each service type provides an internal IP address.
5. (b) – Monitoring Metric Writer
6. (c) – 0.5%
7. (b) – Kaniko cache
8. (a) – Incident Commander
9. (b) – `gcloud auth configure-docker`
10. (b) – Active development
11. (d) – Sensitive information should be stored in Google Secrets Manager.
12. (d) – Verify whether the service account tied to Compute Engine has the required role to send to cloud logging.
13. (b) and (c) – `3xx` status count and `4xx` status count
14. (b) – `spec.template`
15. (b) – There are no consequences for missing an SLO.
16. (c) – Data Access
17. (b) – The Incident Commander should sign off on all decisions during incident management.
18. (a) – SLI
19. (c) – Go to Cloud Error Reporting and view the error logs aggregated by occurrences.
20. (c) – Data Access
21. (b) – Set the SLO based on collecting historical data during the pre-GA period.
22. (c) – Admin activity logs cannot be deleted.
23. (c) – Cloud Build publishes a message to a pub/sub topic. The CD tool is subscribed to the topic.
24. (a) and (d) – Provides enduring value and is tactical
25. (d) – Binary authorization
26. (c) – Once

27. (d) – The retention period is 400 days and is not configurable.

28. (d) – Container analysis

29. (c) – Create an attestation and submit to binary authorization

30. (b) – Service account

31. (d)

32. (b) – Minor

33. (b) – System Event

34. (b) and (d) – Throughput and end-to-end latency

35. (c) – `ClusterIP`

36. (c) – Enable VPC flow logs

37. (b) – Namespace scoped

38. (c) – Containers

39. (c) – Global

40. (d)

41. (b) – Design components to be stateful wherever possible.

42. (b)

43. (b) – Node

44. (b) – Deploy the monitoring application as a `DaemonSet` object.

45. (b) – StatefulSet controller

46. (c) – Cloud logging: Create a log sink and export it to BigQuery.

47. (b) – SLO

48. (d) – None of the above (Private Logs Viewer)

49. (b) – Communications Lead

50. (b) – Create an alerting policy that considers a shorter lookback period and a longer lookback period.

Mock Exam 2

Test Duration: 2 hours

Total Number of Questions: 50

Answer the following questions:

1. Which of the following Google Cloud services is suitable for storing Docker images? Select all appropriate options.

   a) Cloud Source Repositories

   b) Container Registry

   c) Cloud Build

   d) Artifact Registry

2. Company A has decided to deploy a containerized application to a GKE cluster. The GKE cluster is made up of multiple nodes or compute machines. The requirement is to collect detailed metrics with respect to deployment, which includes node- and container-related metrics. Which of the following is the recommended approach to collect these metrics?

   a) Install the Monitoring agent on the GKE cluster.

   b) Enable Cloud Operations on the GKE cluster.

   c) Install the Logging agent on the GKE cluster.

   d) (a) and (b)

3. Which of the following is not a valid option while configuring a private GKE cluster?

   a) Public endpoint access disabled.

   b) Public endpoint access enabled; authorized networks enabled for limited access.

   c) Public endpoint access disabled; authorized networks enabled for limited access.

   d) Public endpoint access enabled; authorized networks disabled.

4. Select the log type that represents log entries tied to setting or changing the permissions of a Cloud Storage bucket:

    a) Admin Activity

    b) System Event

    c) Data Access

    d) Access Transparency

5. Your team has decided to analyze logs in Cloud Logging through BigQuery. Select the option that is most appropriate to achieve the team's goal:

    a) Export to Cloud Storage by creating a log sink, create an event in a Cloud Storage bucket, and push it to BigQuery.

    b) Export to BigQuery by creating a log sink.

    c) Export to Pub/Sub by creating a log sink and configure BigQuery as a subscriber to the topic.

    d) Cannot export to BigQuery.

6. Select two options from the following that Cloud Build does not support as a valid machine type to initiate the Cloud Build process.

    a) M1 – Memory Optimized

    b) N1 – General Purpose

    c) C2 – Compute Optimized

    d) E2 – General Purpose

7. Which of the following is a characteristic of black-box monitoring?

    a) Symptom-oriented and cause-oriented

    b) Collects information from metrics, logs, and traces

    c) Best used for the paging of incidents after the incident has occurred

    d) Metrics exposed by the internals of the system

8. A debug snapshot captures _____.

    a) local variables

    b) the call stack at a specific location

    c) Both (a) and (b)

    d) None of the above

9. _____ logs will record operations that modify the configuration or metadata of Compute Engine resources.

a) Admin Activity

b) System Event

c) Data Access

d) Access Transparency

10. _____ stores trusted metadata used in the authorization process.

a) Container Registry

b) Cloud Storage

c) Container Analysis

d) Binary Authorization

11. As per SRE, a service moves to the General Availability phase when _____.

a) agreed-upon GA data has been reached

b) the Production Readiness Review has passed

c) an executive has used their silver tokens

d) the product head has raised a request to push to production

12. _____ is not a function of Log Router.

a) Ingest Logs

b) Debug Logs

c) Discard Logs

d) Export Logs

13. Which of the following options requires you to install a Logging agent?

a) View linux syslogs

b) View admin activity logs

c) View windows event view logs

d) (a) and (c)

14. _____ logs will record automatic restart operations for Google Compute Engine.

    a) Admin Activity

    b) System Event

    c) Data Access

    d) Access Transparency

15. When creating a GKE cluster in Autopilot mode, select the possible configuration choices from the following options with respect to network isolation:

    a) Private or public cluster

    b) Public cluster only

    c) Private cluster only

    d) None

16. Select the option that is best suited to analyze network traffic at a subnet level in a VPC:

    a) Data Access Logs

    b) Flow Logs

    c) Admin Logs

    d) Firewall Logs

17. _____ logs will record operations changing any properties of a resource such as tags/labels for Google Compute Engine.

    a) Admin Activity

    b) System Event

    c) Data Access

    d) Access Transparency

18. An application accepts inputs from users and needs to initiate background tasks accordingly to perform automated asynchronous execution. Select the appropriate Google Cloud service:

    a) Cloud Pub/Sub

    b) Cloud Task

    c) Cloud Scheduler

    d) Cloud Cron

19. Container Analysis stores trusted metadata using _____. This metadata is used in the binary authorization process.

    a) Cloud Storage

    b) Container Registry

    c) Cloud Source Repositories

    d) Persistent Disk or Persistent Volume

20. Select the compute option that is the most managed out of the following:

    a) Google Compute Engine

    b) Google App Engine

    c) Google Kubernetes Engine

    d) Google Cloud Run

21. _____ logs will record operations of setting or changing permissions for Compute Engine.

    a) Admin Activity

    b) System Event

    c) Data Access

    d) Access Transparency

22. Which of the following is true with respect to the default retention period for user logs stored in the default log bucket?

    a) It's 30 days and it cannot be changed.

    b) It's 400 days and it can be changed from between 1 day and 10 years.

    c) It's 400 days and cannot be changed.

    d) It's 30 days and can be changed to between 1 day and 10 years.

23. Select the order of Google Cloud services to implement CI/CD:

    a) Cloud Build, Cloud Source Repositories, Container Registry, GKE

    b) Cloud Source Repositories, Cloud Build, Container Registry, GKE

    c) Cloud Source Repositories, Container Registry, Cloud Build, GKE

    d) Cloud Build, Container Registry, Cloud Source Repositories, GKE

24. Select the option to retain Admin Activity logs beyond the retention period:

    a) Admin Activity logs never expire.

    b) Admin Activity logs cannot be retained beyond the retention period.

    c) Admin Activity logs can be exported to Cloud Storage.

    d) Admin Activity logs that have expired can be recovered through the CLI.

25. Select two statements that are true with respect to toil in SRE:

    a) Toil is used to track error budgets.

    b) Reducing toil is a critical task of an SRE engineer.

    c) Toil is a repetitive task that is not tied to the production system.

    d) Toil is a repetitive task that is tied to the production system.

26. Select the most suitable definition for a StatefulSet:

    a) Represents a set of services with unique, persistent identities and stable hostnames

    b) Represents a set of clusters with unique, persistent identities and stable hostnames

    c) Represents a set of pods with unique, persistent identities and stable hostnames

    d) Represents a set of Docker images with unique, persistent identities and stable hostnames

27. Cloud Trace can collect latency data from _____.

    a) load balancers

    b) GKE applications

    c) Both

    d) None

28. The goal is to create a GKE cluster where nodes are configured to support resiliency and high availability without manual intervention. Which of the following GKE features is most appropriate?

    a) GKE's node auto-upgrade feature

    b) GKE's node auto-repairing feature

    c) GKE's node auto-healing feature

    d) Both (a) and (c)

29. Which of the following is used for storing records related to administrative events?

    a) Tracing

    b) Logging

    c) Error reporting

    d) Debugging

30. An application is deployed using Deployment Manager. You need to update the deployment with minimal downtime. Select the appropriate command:

    a) `gcloud deployment-manager deployments update`

    b) `gcloud deployment manager deployment update`

    c) `gcloud deployment manager deployments update`

    d) `gcloud deployment-manager deployment update`

31. An application deployed on Google Cloud Platform is having issues with respect to latency. Select the Google Cloud service that can inspect latency data in near real time:

    a) Cloud Networking

    b) Cloud Trace

    c) Cloud Profiler

    d) Cloud Debugger

32. Select the best practices to conduct a blameless postmortem (select multiple):

    a) Provide clarity that helps in future mitigation.

    b) Generate an incident report.

    c) Outline the events of the incident.

    d) Identify the team members responsible for the incident.

33. Select the feature supported by GKE to increase cluster security using a verifiable node identity:

    a) Workload identity

    b) Shielded GKE nodes

    c) Binary authorization

    d) (a) and (b)

34. Which of the following is not a valid option when trying to connect an AWS account to a workspace?

    a) A GCP connector project is required.

    b) A GCP connector project needs to be in a different parent organization than the workspace.

    c) A GCP connector project needs to be in the same parent organization as the workspace.

    d) The billing account should be tied to the connector project.

35. You are tasked to choose SLIs for a user-facing system. Which of the following are not part of the four golden signals as per Google Cloud best practices (select multiple)?

    a) Saturation

    b) Throughput

    c) Traffic

    d) Correctness

36. Select the best option to handle unexpected surges in traffic to a GKE cluster:

    a) Create two separate clusters of different sizes. If the traffic increases, send traffic to the bigger cluster.

    b) Put a cap on the amount of traffic that your GKE cluster can handle.

    c) Enable autoscaling on the GKE cluster.

    d) Increase the size of the cluster or change the machine type of the nodes in the cluster.

37. Based on the principle of least privilege, select the IAM role that allows an end user to create or modify an export sink from Cloud Logging:

    a) Project Editor

    b) Logging Admin

    c) Logs Configuration Writer

    d) Project Viewer

38. Which of the following is the first step with respect to Cloud Deployment Manager?

    a) Create a resource.

    b) Create a template.

    c) Create a configuration.

    d) Create a deployment.

39. An application runs on GKE. There is a desire to use Spinnaker to perform blue/green deployments. Select the most appropriate option:

    a) Use a Kubernetes Deployment and then use Spinnaker to update the Deployment for each new version deployed.

    b) Use a Kubernetes ReplicaSet and then use Spinnaker to update the ReplicaSet for each new version deployed.

    c) Use a Kubernetes DaemonSet and then use Spinnaker to update the DaemonSet for each new version deployed.

    d) Use a Kubernetes StatefulSet and then use Spinnaker to update the StatefulSet for each new version deployed.

40. You need to create a separate namespace for UAT. This is a way to isolate the deployment. Select the most appropriate command:

    a) `kubectl namespace create uat`

    b) `kubectl create namespace uat`

    c) `kubectl namespace uat create`

    d) `kubectl create uat namespace`

41. Cloud Deployment Manager allows you to specify the resources required for your application in a declarative format using YAML. Which flag allows you to verify the resources that could be created before doing the actual implementation?

    a) `--dry-run`

    b) `--preview`

    c) `--snapshot`

    d) `--verify`

42. _____ is not considered as a Kubernetes workload resource.

    a) A Deployment

    b) A Service

    c) A Cronjob

    d) A DaemonSet

43. A resource in Cloud Deployment Manager represents _____.

    a) a single API resource

    b) a compute resource

    c) a cloud database resource

    d) a Cloud Storage resource

44. Which of the following commands is used to install `kubectl` on Google Cloud Shell?

    a) `gcloud components install kubectl`

    b) `gcloud component kubectl install`

    c) `gcloud components kubectl install`

    d) `gcloud component install kubectl`

45. While creating a cluster in standard mode, the maximum pods per node defaults to _____.

    a) 115

    b) 100

    c) 110

    d) 105

46. Select the command that can manually increase the number of nodes in a GKE cluster:

    a) `gcloud container clusters resize`

    b) `kubectl containers clusters resize`

    c) `kubectl container clusters resize`

    d) `gcloud containers clusters resize`

47. The goal is to store audit logs for long-term access and allow access to external auditors. Which of the following are the most appropriate steps (select two)?

    a) Export audit logs to Cloud Storage by creating a log sink.

    b) Export audit logs to BigQuery by creating a log sink.

    c) Assign the IAM role to auditors so that they have access to the BigQuery dataset that contains the audit logs.

    d) Assign the IAM role to auditors so that they have access to the storage bucket that contains the audit logs.

48. A user is trying to deploy an application to a GKE cluster. The user is using `kubectl` to execute some commands but `kubectl` is unable to connect with the cluster's `kube-apiserver`. Which of the following is the possible reason for this issue and the potential fix?

    a) The firewall ingress rule is not configured correctly, blocking incoming traffic from the user to the cluster.

    b) `kubeconfig` is missing the credentials to connect and authenticate with the cluster. Run the `gcloud container clusters auth login` command.

    c) The firewall egress rule is not configured correctly, blocking outgoing traffic from the cluster to the user.

    d) `kubeconfig` is missing the credentials to connect and authenticate with the cluster. Run the `gcloud container clusters get-credentials` command.

49. _____ is a GCP service that is used for infrastructure automation.

    a) Terraform

    b) Cloud Terraform

    c) Cloud Puppet

    d) Cloud Deployment Manager

50. GKE allows up to a maximum of _____ clusters per zone.

    a) 50

    b) 25

    c) 55

    d) 40

Answers

1. (b) and (d) – Container Registry and Artifact Registry.

2. (b) – Enable Cloud Operations on the GKE cluster.

3. (c) – Public endpoint access disabled; authorized networks enabled for limited access.

4. (a) – Admin Activity

5. (b) – Export to BigQuery by creating a log sink.

6. (a) and (c) – M1 – Memory Optimized and C2 – Compute Optimized

7. (c) – Best used for the paging of incidents after the incident has occurred

8. (c) – Both (a) and (b)

9. (a) – Admin Activity

10. (c) – Container Analysis

11. (b) – the Production Readiness Review has passed

12. (b) – Debug Logs

13. (d) – (a) and (c)

14. (b) – System Event

15. (a) – Private or public cluster

16. (b) – Flow Logs

17. (a) – Admin Activity

18. (b) – Cloud Task

19. (b) – Container Registry

20. (d) – Google Cloud Run

21. (a) – Admin Activity

22. (d) – It's 30 days and can be changed to between 1 day and 10 years.

23. (b) – Cloud Source Repositories, Cloud Build, Container Registry, GKE

24. (c) – Admin Activity logs can be exported to Cloud Storage.

25. (b) and (d) – Reducing toil is a critical task of an SRE engineer and Toil is a repetitive task that is tied to the production system.

26. (c) – Represents a set of pods with unique, persistent identities and stable hostnames

27. (c) – Both

28. (b) – GKE's node auto-repairing feature

29. (b) – Logging

30. (b) – `gcloud deployment manager deployment update`

31. (b) – Cloud Trace

32. (a), (b), and (c) – Provide clarity that helps in future mitigation, Generate an incident report, and Outline the events of the incident.

33. (b) – Shielded GKE nodes

34. (b) – A GCP connector project needs to be in a different parent organization than the workspace.

35. (b) and (d) – Throughput and Correctness (not part of the 4 golden signals – Latency, Error, Traffic, Saturation)

36. (c) – Enable autoscaling on the GKE cluster.

37. (c) – Logs Configuration Writer

38. (c) – Create a configuration.

39. (b) – Use a Kubernetes ReplicaSet and then use Spinnaker to update the ReplicaSet for each new version deployed.

40. (b) – `kubectl create namespace uat`

41. (b) – `--preview`

42. (b) – A Service

43. (a) – a single API resource

44. (a) – `gcloud components install kubectl`

45. (c) – 110

46. (a) – `gcloud container clusters resize`

47. (a) and (d) – Export audit logs to Cloud Storage by creating a log sink and Assign the IAM role to auditors so that they have access to the audit bucket that contains the storage logs.

48. (b) – kubeconfig is missing the credentials to connect and authenticate with the cluster. Run the gcloud container clusters auth login command.

49. (d) – Cloud Deployment Manager

50. (a) – 50

Packt.com

Subscribe to our online digital library for full access to over 7,000 books and videos, as well as industry leading tools to help you plan your personal development and advance your career. For more information, please visit our website.

Why subscribe?

- Spend less time learning and more time coding with practical eBooks and Videos from over 4,000 industry professionals

- Improve your learning with Skill Plans built especially for you

- Get a free eBook or video every month

- Fully searchable for easy access to vital information

- Copy and paste, print, and bookmark content

Did you know that Packt offers eBook versions of every book published, with PDF and ePub files available? You can upgrade to the eBook version at packt.com and as a print book customer, you are entitled to a discount on the eBook copy. Get in touch with us at customercare@packtpub.com for more details.

At www.packt.com, you can also read a collection of free technical articles, sign up for a range of free newsletters, and receive exclusive discounts and offers on Packt books and eBooks.

Other Books You May Enjoy

If you enjoyed this book, you may be interested in these other books by Packt:

Professional Cloud Architect – Google Cloud Certification Guide

Konrad Cłapa , Brian Gerrard

ISBN: 978-1-83855-527-6

- Manage your GCP infrastructure with Google Cloud management options such as CloudShell and SDK
- Understand the use cases for different storage options
- Design a solution with security and compliance in mind
- Monitor GCP compute options
- Discover machine learning and the different machine learning models offered by GCP
- Understand what services need to be used when planning and designing your architecture

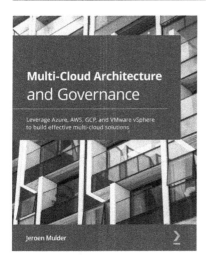

Multi-Cloud Architecture and Governance

Jeroen Mulder

ISBN: 978-1-80020-319-8

- Get to grips with the core functions of multiple cloud platforms
- Deploy, automate, and secure different cloud solutions
- Design network strategy and get to grips with identity and access management for multi-cloud
- Design a landing zone spanning multiple cloud platforms
- Use automation, monitoring, and management tools for multi-cloud
- Understand multi-cloud management with the principles of BaseOps, FinOps, SecOps, and DevOps
- Define multi-cloud security policies and use cloud security tools
- Test, integrate, deploy, and release using multi-cloud CI/CD pipelines

Packt is searching for authors like you

If you're interested in becoming an author for Packt, please visit authors.
packtpub.com and apply today. We have worked with thousands of developers and
tech professionals, just like you, to help them share their insight with the global tech
community. You can make a general application, apply for a specific hot topic that we
are recruiting an author for, or submit your own idea.

Leave a review - let other readers know what you think

Please share your thoughts on this book with others by leaving a review on the site that
you bought it from. If you purchased the book from Amazon, please leave us an honest
review on this book's Amazon page. This is vital so that other potential readers can see
and use your unbiased opinion to make purchasing decisions, we can understand what
our customers think about our products, and our authors can see your feedback on the
title that they have worked with Packt to create. It will only take a few minutes of your
time, but is valuable to other potential customers, our authors, and Packt. Thank you!

Index

A

time series
  about 87, 94
  cardinality 97
  structure 95, 96
time series, metric types
  counter 97
  distribution 98
  gauge 97
Time to detect (TTD)
  about 62
  reducing 63
Time to fail (TTF) 62
Time to resolve (TTR)
  about 62
  reducing 63
toil
  about 12, 19
  advantages 65
  removing, through automation 65, 66
tools SRE team 107
total events 49
trace 378
Trace List window 380, 381
Trace Overview page 379
traces 85
triggers
  used, for invoking manual build
    process through Cloud Build 172

U

unified vision 125
universal code search
  performing 151, 152

uptime check, Cloud Monitoring 347-349
uptime check failures
  potential reasons 349
user-defined (logs-based) metrics 367
user interactions
  bucketizing 49
user journey
  about 13
  categorizing 47
  correctness 50
  data processing/pipeline-
    based user journey 50
  request/response user journey 48

V

Vertical Pod Autoscaler (VPA) 277, 280
virtualization 162
virtual machines (VMs) 207
Virtual Private Cloud (VPC) 281
virtual Trusted Platform
    Modules (vTPMs) 324
Visual Studio Code 27
VPC flow logs 368
VPC-native cluster 282

W

well-defined SLA
  blueprint 39, 40
white box monitoring 84, 85
Workload Identity
  about 328, 329
  enabling 329

www.ingramcontent.com/pod-product-compliance
Lightning Source LLC
Chambersburg PA
CBHW081455050326
40690CB00015B/2811